THE EQUATIONS AND SOLUTIONS TO UNSOLVED PROBLEMS

——THE EQUATIONS——

THE PINNACLE EQUATIONS p. 5

SOLUTIONS TO LONG-STANDING PROBLEMS
Biology, p. 33
Chemistry, p. 59
Computer Science, p. 67
Environmentalism, p. 84
Humanitarianism, p. 89
Mathematics, p. 90
Neurology, p. 144
Philosophy, p. 155
Physics, p. 185
Society, p. 237

MISCELLANEOUS EQUATIONS p. 241

PSEUDO EQUATIONS p. 283

STATE OF KNOWLEDGE, p. 287

PRECEDENTS p. 289

FURTHER READING p. 303

BIO p. 304

Nathan Coppedge

© 2018, 2019, 2022 Nathan Coppedge. Some materials previously released under copyright for re-use with citation of the author.

THE EQUATIONS AND SOLUTIONS TO UNSOLVED PROBLEMS

Expanded Edition.

BY NATHAN COPPEDGE

Nathan Coppedge

,,,

4

THE EQUATIONS AND SOLUTIONS TO UNSOLVED PROBLEMS

PINNACLE EQUATIONS

A Coherent Set of Equations.

The goal here is to repeat the types of equations found in the perpetual motion mathematics for each variation of the Pinnacle Theory Model with a view to exp eff.

LOOP 1:

Efficiency - Difference (= Disintegral / Set theory)

- Min Results = (Max Eff / 2) - Diff
- Max Results = Min Eff - Diff
- Min Eff = Results + Diff
- Max Eff = (Min Results + Diff) X 2
- Min Difference = - (Results - Max Efficiency)
- Max Difference = - (Results - Min Efficiency)
- Over-Unity = ((Max Results - Min Results) / (Max Eff)) - Diff X 100 (%)
- Proportion of Smaller Unit of Disintegral = 1X

- Efficiency - Difference (= Negative Disintegral) Keep items on right of equals related to disintegral not negative disintegral

- Min Results = - [(Max Eff / 2) - Diff]
- Max Results = -(Min Eff - Diff)
- Min Eff = -(Results + Diff)
- Max Eff = -[(Min Results + Diff) X 2]
- Min Difference = (Results - Max Efficiency) Not negative for negative disintegral
- Max Difference = (Results - Min Efficiency) Not negative for negative disintegral
- Over-Unity = - [((Max Results - Min Results) / (Max Eff)) - Diff] X 100 (%)
- Proportion of Smaller Unit of Negative Disintegral = 1X

Difference - Efficiency (= Antitheory)

- Min Results = Diff - (Max Eff / 2)
- Max Results = Diff - Min Eff
- Min Eff = Diff - Results
- Max Eff = (Diff - Min Results) X 2
- Min Difference = Results + Max Efficiency
- Max Difference = Results + Min Efficiency
- Over-Unity = ((Max Results - Min Results) / -(Max Eff)) + Diff X 100 (%)
- Proportion of Smaller Unit = 1X

Efficiency + Difference (= TOE)

- Min Results = (Max Eff / 2) + Diff
- Max Results = Min Eff + Diff
- Min Eff = Results - Diff
- Max Eff = (Min Results - Diff) X 2
- Min Difference = Results - Max Efficiency
- Max Difference = Results - Min Efficiency
- Over-Unity = ((Max Results - Min Results) / (Max Eff)) + Diff X 100 (%)
- Proportion of Smaller Unit = 1X

THE EQUATIONS AND SOLUTIONS TO UNSOLVED PROBLEMS

LOOP 2:

Difference + Results (= Efficiency Disintegral)

- Min Efficiency = (Max Results / 2) + Diff
- Max Efficiency = Min Results + Diff
- Min Results = Eff - Diff
- Max Results = (Min Eff - Diff) X 2
- Min Difference = Eff - Max Results
- Max Difference = Eff - Min Results
- Over-Unity = ((Max Eff - Min Eff) / (Max Results)) + Diff X 100 (%)
- Proportion of Smaller Unit = 1X

Results - Difference (= Efficiency)

- Min Efficiency = (Max Results / 2) - Diff
- Max Efficiency = Min Results - Diff
- Min Results = Eff + Diff
- Max Results = (Min Eff + Diff) X 2
- - (Min Difference) = Eff - Max Results
- - (Max Difference) = Eff - Min Results
- Over-Unity = ((Max Eff - Min Eff) / (Max Results)) - Diff X 100 (%)
- Proportion of Smaller Unit = 1X

-Difference - Results (= Anti-Efficiency Disintegral, different equation than efficiency disintegral in this case, simply negated, an alternate equation is that it is non-negated)

- Min Efficiency = - [(Max Results / 2) + Diff]
- Max Efficiency = - (Min Results + Diff)

- Min Results = -(Eff - Diff)
- Max Results = - [(Min Eff - Diff) X 2]
- Min Difference = - (Eff - Max Results)
- Max Difference = - (Eff - Min Results)
- Over-Unity = - [((Max Eff - Min Eff) / (Max Results)) + Diff] X 100 (%)
- Proportion of Smaller Unit = 1X

-Results + Difference (= Anti-Efficiency)

- Min Efficiency = (- Max Results / 2) + Diff
- Max Efficiency = (- Min Results) + Diff
- - (Min Results) = Eff - Diff
- - (Max Results) = (Min Eff - Diff) X 2
- Min Difference = Eff + Max Results
- Max Difference = Eff + Min Results
- Over-Unity = ((Max Eff - Min Eff) / (- Max Results)) + Diff X 100 (%)
- Proportion of Smaller Unit = 1X

LOOP 3

Results + Difference (= Anti-Efficiency Dis-integral, same equations in this case as efficiency disintegral, but non-overlapping)

- Min Efficiency = (Max Results / 2) + Diff
- Max Efficiency = Min Results + Diff
- Min Results = Eff - Diff
- Max Results = (Min Eff - Diff) X 2
- Min Difference = Eff - Max Results
- Max Difference = Eff - Min Results

THE EQUATIONS AND SOLUTIONS TO UNSOLVED PROBLEMS

- Over-Unity = ((Max Eff - Min Eff) / (Max Results)) + Diff X 100 (%)
- Proportion of Smaller Unit = 1X Difference - Results (= Anti-Efficiency)

- Min Efficiency = (- (Max Results) / 2) + Diff
- Max Efficiency = - (Min Results) + Diff
- - (Min Results) = Eff - Diff
- - (Max Results) = (Min Eff - Diff) X 2
- Min Difference = Eff + Max Results
- Max Difference = Eff + Min Results
- Over-Unity = ((Max Eff - Min Eff) / (-Max Results)) + Diff X 100 (%)
- Proportion of Smaller Unit = 1X Results + Difference (= Anti-Efficiency Dis-integral, same as anti-efficiency disintegral)

- Min Efficiency = (Max Results / 2) + Diff
- Max Efficiency = Min Results + Diff
- Min Results = Eff - Diff
- Max Results = (Min Eff - Diff) X 2
- Min Difference = Eff - Max Results
- Max Difference = Eff - Min Results
- Over-Unity = ((Max Eff - Min Eff) / (Max Results)) + Diff X 100 (%)
- Proportion of Smaller Unit = 1X Difference - Results (= Anti-Efficiency, same as earlier)

- Min Efficiency = (- (Max Results) / 2) + Diff
- Max Efficiency = - (Min Results) + Diff
- - (Min Results) = Eff - Diff

- - (Max Results) = (Min Eff - Diff) X 2
- Min Difference = Eff + Max Results
- Max Difference = Eff + Min Results
- Over-Unity = ((Max Eff - Min Eff) / (- Max Results)) + Diff X 100 (%)
- Proportion of Smaller Unit = 1X

LOOP 4

Efficiency + Results (= Difference Disintegral, Note: Because the Results and Difference are swapped in this equation, the difference has been intentionally listed earlier in the list of equations in this specific case)

- Min Difference = (Max Eff / 2) + Results
- Max Difference = Min Eff + Results
- Min Eff = Difference - Results
- Max Eff = (Min Diff - Results) X 2
- Min Results = Difference - Max Efficiency
- Max Results = Difference - Min Efficiency
- Over-Unity = ((Max Difference - Min Difference) / (Max Eff)) + Results X 100 (%)
- Proportion of Smaller Unit = 1X

Results – Efficiency (= Difference)

- Min Difference = (- Max Eff / 2) + Results
- Max Difference = (- Min Eff) + Results
- - (Min Eff) = Difference - Results
- - (Max Eff) = (Min Diff - Results) X 2

THE EQUATIONS AND SOLUTIONS TO UNSOLVED PROBLEMS

- Min Results = Difference + Max Efficiency
- Max Results = Difference + Min Efficiency
- Over-Unity = ((Max Difference - Min Difference) / (- Max Eff)) + Results X 100 (%)
- Proportion of Smaller Unit = 1X

-Efficiency - Results (= Anti-Difference Disintegral, this uses difference earlier like difference disintegral but is negated)

- Min Difference = - [(Max Eff / 2) + Results]
- Max Difference = - (Min Eff + Results)
- Min Eff = - (Difference - Results)
- Max Eff = - [(Min Diff - Results) X 2]
- Min Results = - (Difference - Max Efficiency)
- Max Results = - (Difference - Min Efficiency)
- Over-Unity = - [((Max Difference - Min Difference) / (Max Eff)) + Results] X 100 (%)
- Proportion of Smaller Unit = 1X

-Results + Efficiency (= Anti-Difference)

- Min Difference = (Max Eff / 2) - Results
- Max Difference = Min Eff - Results
- Min Eff = Difference + Results
- Max Eff = (Min Diff + Results) X 2
- - (Min Results) = Difference - Max Efficiency
- - (Max Results) = Difference - Min Efficiency
- Over-Unity = ((Max Difference - Min

11

Difference) / (Max Eff)) - Results X 100 (%)
- Proportion of Smaller Unit = 1X

LOOP 5

Antiforces + # Dimensions (= Force Dis-integral)

- Min Forces = (Dimensions / 2) + Antiforces
- Max Forces = Min Dimensions + Antiforces
- Min Dimensions = Forces - Antiforces
- Max Dimensions = (Min Forces - Antiforces) X 2
- Min Antiforces = Forces - Max Dimensions
- Max Antiforces = Forces - Min Dimensions
- Over-Unity = ((Max Forces - Min Forces) / (Max Dimensions)) + Antiforces X 100 (%)
- Proportion of Smaller Unit = 1X

Dimensions - #Antiforces (= Forces)

- Min Forces = (Dimensions / 2) - Antiforces
- Max Forces = Min Dimensions - Antiforces
- Min Dimensions = Forces + Antiforces
- Max Dimensions = (Min Forces + Antiforces) X 2
- - (Min Antiforces) = Forces - Max Dimensions
- - (Max Antiforces) = Forces - Min Di-

THE EQUATIONS AND SOLUTIONS TO UNSOLVED PROBLEMS

mensions
- Over-Unity = ((Max Forces - Min Forces) / (Max Dimensions)) - Antiforces X 100 (%)
- Proportion of Smaller Unit = 1X

-#Antiforces - #Dimensions (= Negative Force Disintegral)

- Min Forces = - [(Dimensions / 2) + Antiforces]
- Max Forces = - (Min Dimensions + Antiforces)
- Min Dimensions = - (Forces - Antiforces)
- Max Dimensions = - [(Min Forces - Antiforces) X 2]
- Min Antiforces = - (Forces - Max Dimensions)
- Max Antiforces = - (Forces - Min Dimensions)
- Over-Unity = - [((Max Forces - Min Forces) / (Max Dimensions)) + Antiforces] X 100 (%)
- Proportion of Smaller Unit = 1X

-#Dimensions + # Antiforces (= Negative Forces)

- Min Forces = (- Dimensions / 2) + Antiforces
- Max Forces = - (Min Dimensions) + Antiforces
- - (Min Dimensions) = Forces - Antiforces
- - (Max Dimensions) = (Min Forces - Antiforces) X 2
- Min Antiforces = Forces + Max Di-

mensions
- Max Antiforces = Forces + Min Dimensions
- Over-Unity = ((Max Forces - Min Forces) / (- Max Dimensions)) + Antiforces X 100 (%)
- Proportion of Smaller Unit = 1X

LOOP 6

\# Forces + #Dimensions (= Antiforce Disintegral)

- Min Antiforces = (Forces / 2) + Dimensions
- Max Antiforces = Min Forces + Dimensions
- Min Forces = Antiforces - Dimensions
- Max Forces = (Min Antiforces - Dimensions) X 2
- Min Dimensions = Antiforces - Max Forces
- Max Dimensions = Antiforces - Min Forces
- Over-Unity = ((Max Antiforces - Min Antiforces) / (Max Forces)) + Dimensions X 100 (%)
- Proportion of Smaller Unit = 1X

\# Dimensions - # Forces (= Antiforces)

- Min Antiforces = (- Forces / 2) + Dimensions
- Max Antiforces = - (Min Forces) + Dimensions
- - (Min Forces) = Antiforces - Dimensions

THE EQUATIONS AND SOLUTIONS TO UNSOLVED PROBLEMS

- - (Max Forces) = (Min Antiforces - Dimensions) X 2
- Min Dimensions = Antiforces + Max Forces, note plus here is intentional
- Max Dimensions = Antiforces + Min Forces, note plus here is intentional
- Over-Unity = ((Max Antiforces - Min Antiforces) / - (Max Forces)) + Dimensions X 100 (%)
- Proportion of Smaller Unit = 1X -#Forces - #Dimensions (= Negative Antiforce Disintegral, this is the negation of the antiforce disintegral)

- Min Antiforces = - [(Forces / 2) + Dimensions]
- Max Antiforces = - (Min Forces + Dimensions)
- Min Forces = - (Antiforces - Dimensions)
- Max Forces = - [(Min Antiforces - Dimensions) X 2]
- Min Dimensions = - (Antiforces - Max Forces)
- Max Dimensions = - (Antiforces - Min Forces)
- Over-Unity = - [((Max Antiforces - Min Antiforces) / (Max Forces)) + Dimensions] X 100 (%)
- Proportion of Smaller Unit = 1X -#Dimensions + #Forces (= Negative Antiforces)

- Min Antiforces = (Forces / 2) - (Dimensions), negative dimensions is intentional

- Max Antiforces = Min Forces - (Dimensions), negative dimensions is intentional
- Min Forces = Antiforces + (Dimensions), positive dimensions is intentional
- Max Forces = (Min Antiforces + Dimensions) X 2, positive dimensions is intentional
- - (Min Dimensions) = Antiforces - Max Forces, remember neg min dimensions
- - (Max Dimensions) = Antiforces - Min Forces, remember neg max dimensions
- Over-Unity = ((Max Antiforces - Min Antiforces) / (Max Forces)) - Dimensions X 100 (%), negative dimensions is intentional
- Proportion of Smaller Unit = 1X

LOOP 7

#Forces - # Antiforces (= Force Dimensions, Dimensional Disintegral)

- Min Dimensions = (Max Forces / 2) - Antiforces
- Max Dimensions = Min Forces - Antiforces
- Min Forces = Dimensions + Antiforces
- Max Forces = (Min Dimensions + Antiforces) X 2
- Min Antiforces = - (Dimensions - Max Forces)
- Max Antiforces = - (Dimensions - Min Forces)

THE EQUATIONS AND SOLUTIONS TO UNSOLVED PROBLEMS

- Over-Unity = ((Max Dimensions - Min Dimensions) / (Max Forces)) - Antiforces X 100 (%)
- Proportion of Smaller Unit of Dimensions = 1X

#Antiforces + #Forces (= Dimensions)

- Min Dimensions = (Forces / 2) + Antiforces
- Max Dimensions = Min Forces + Antiforces
- Min Forces = Dimensions - Antiforces
- Max Forces = (Min Dimensions - Antiforces) X 2
- Min Antiforces = Dimensions - Max Forces
- Max Antiforces = Dimensions - Min Forces
- Over-Unity = ((Max Dimensions - Min Dimensions) / (Max Forces)) + Antiforces X 100 (%)
- Proportion of Smaller Unit = 1X

-#Forces + #Antiforces (= Antiforce Dimensions, Hyper-Dimensional Disintegral)

- Min Dimensions = Antiforces - (Max Forces / 2)
- Max Dimensions = Antiforces - Min Forces
- Min Forces = Antiforces - Dimensions
- Max Forces = (Antiforces - Min Dimensions) X 2
- Min Antiforces = Dimensions + Max Forces
- Max Antiforces = Dimensions + Min Forces

- Over-Unity = ((Max Dimensions - Min Dimensions) / -(Max Forces)) + Anti-forces X 100 (%)
- Proportion of Smaller Unit = 1X

-#Antiforces -#Forces (= Hyper-Dimensions, may translate as dimensions due to using negative units, alternately similar to negative dimensions)

- Min Hyper-Dimensions = - [(Forces / 2) + Antiforces]
- Max Hyper-Dimensions = - [Min Forces + Antiforces]
- Min Forces = - [Dimensions - Antiforces]
- Max Forces = - [(Min Dimensions - Antiforces) X 2]
- Min Antiforces = - [Dimensions - Max Forces]
- Max Antiforces = - [Dimensions - Min Forces]
- Over-Unity = -[((Max Dimensions - Min Dimensions) / (Max Forces)) + Antiforces] X 100 (%)
- Proportion of Smaller Unit = 1X

LOOP 8

-#Forces - #Antiforces (= Anti-Hyper-Dimensions same as hyper-d, Anti-Dimensional Disintegral, may translate as dimensions due to using negative units, alternately similar to negative dimensions)

- Min Hyper-Dimensions = - [(Forces / 2) + Antiforces]

THE EQUATIONS AND SOLUTIONS TO UNSOLVED PROBLEMS

- Max Hyper-Dimensions = - [Min Forces + Antiforces]
- Min Forces = - [Dimensions - Antiforces]
- Max Forces = - [(Min Dimensions - Antiforces) X 2]
- Min Antiforces = - [Dimensions - Max Forces]
- Max Antiforces = - [Dimensions - Min Forces]
- Over-Unity = -[((Max Dimensions - Min Dimensions) / (Max Forces)) + Antiforces] X 100 (%)
- Proportion of Smaller Unit = 1X

Antiforces - # Forces (= Anti-Dimensions, same equations as Antiforce Dimensions, but non-overlapping)

- Min Dimensions = Antiforces - (Max Forces / 2)
- Max Dimensions = Antiforces - Min Forces
- Min Forces = Antiforces - Dimensions
- Max Forces = (Antiforces - Min Dimensions) X 2
- Min Antiforces = Dimensions + Max Forces
- Max Antiforces = Dimensions + Min Forces
- Over-Unity = ((Max Dimensions - Min Dimensions) / -(Max Forces)) + Antiforces X 100 (%)
- Proportion of Smaller Unit = 1X

#Forces + #Antiforces (= Dimensions)

- Min Dimensions = (Forces / 2) + Anti-

forces
- Max Dimensions = Min Forces + Antiforces
- Min Forces = Dimensions - Antiforces
- Max Forces = (Min Dimensions - Antiforces) X 2
- Min Antiforces = Dimensions - Max Forces
- Max Antiforces = Dimensions - Min Forces
- Over-Unity = ((Max Dimensions - Min Dimensions) / (Max Forces)) + Antiforces X 100 (%)
- Proportion of Smaller Unit = 1X

-#Antiforces + #Forces (= Force Dimensions)

MIRROR IMAGE OF LOOP 7

- Min Dimensions = (Max Forces / 2) - Antiforces
- Max Dimensions = Min Forces - Antiforces
- Min Forces = Dimensions + Antiforces
- Max Forces = (Min Dimensions + Antiforces) X 2
- Min Antiforces = - (Dimensions - Max Forces)
- Max Antiforces = - (Dimensions - Min Forces)
- Over-Unity = ((Max Dimensions - Min Dimensions) / (Max Forces)) - Antiforces X 100 (%)
- Proportion of Smaller Unit of Dimensions = 1X

THE EQUATIONS AND SOLUTIONS TO UNSOLVED PROBLEMS

LOOP 9

Difference − Efficiency (= Anti-Disintegral / Antitheory)

- Min Results = Diff - (Max Eff / 2)
- Max Results = Diff - Min Eff
- Min Eff = Diff - Results
- Max Eff = (Diff - Min Results) X 2
- Min Difference = Results + Max Efficiency
- Max Difference = Results + Min Efficiency
- Over-Unity = ((Max Results - Min Results) / -(Max Eff)) + Diff X 100 (%)
- Proportion of Smaller Unit = 1X

Efficiency + Difference (= TOE)

- Min Results = (Max Eff / 2) + Diff
- Max Results = Min Eff + Diff
- Min Eff = Results - Diff
- Max Eff = (Min Results - Diff) X 2
- Min Difference = Results - Max Efficiency
- Max Difference = Results - Min Efficiency
- Over-Unity = ((Max Results - Min Results) / (Max Eff)) + Diff X 100 (%)
- Proportion of Smaller Unit = 1X

-Difference + Efficiency (= Disintegral)

- Min Disintegral Results = (Max Disintegral Eff / 2) - Disintegral Diff
- Max Disintegral Results = Min Disintegral Eff - Disintegral Diff
- Min Disintegral Eff = Disintegral Re-

sults + Disintegral Diff
- Max Disintegral Eff = (Min Disintegral Results + Disintegral Diff) X 2
- Min Disintegral Difference = - (Disintegral Results - Max Disintegral Efficiency)
- Max Disintegral Difference = - (Disintegral Results - Min Disintegral Efficiency)
- Disintegral Over-Unity = ((Max Disintegral Results - Min Disintegral Results) / (Max Disintegral Eff)) - Disintegral Diff X 100 (%)
- Proportion of Smaller Unit of Disintegral = 1X

-Efficiency - Difference (= Negative Disintegral)

- Min Negative Disintegral Results = - [(Max Disintegral Eff / 2) - Disintegral Diff]
- Max Negative Disintegral Results = - (Min Disintegral Eff - Disintegral Diff)
- Min Negative Disintegral Eff = - (Disintegral Results + Disintegral Diff)
- Max Negative Disintegral Eff = -[(Min Disintegral Results + Disintegral Diff) X 2]
- Min Negative Disintegral Difference = (Disintegral Results - Max Disintegral Efficiency) Not negative for negative disintegral
- Max Negative Disintegral Difference = (Disintegral Results - Min Disintegral Efficiency) Not negative for negative disintegral
- Negative Disintegral Over-Unity = -

THE EQUATIONS AND SOLUTIONS TO UNSOLVED PROBLEMS

[((Max Disintegral Results - Min Disintegral Results) / (Max Disintegral Eff)) - Disintegral Diff] X 100 (%)
- Proportion of Smaller Unit of Negative Disintegral = 1X

LOOP 10

-Inf Difference + Inf Efficiency (= Super-Disintegral)

- Min Inf Disintegral Results = (Max Inf Disintegral Eff / 2) - Inf Disintegral Diff
- Max Inf Disintegral Results = Min Inf Disintegral Eff - Inf Disintegral Diff
- Min Inf Disintegral Eff = Inf Disintegral Results + Inf Disintegral Diff
- Max Inf Disintegral Eff = (Min Inf Disintegral Results + Inf Disintegral Diff) X 2
- Min Inf Disintegral Difference = - (Inf Disintegral Results - Max Inf Disintegral Efficiency)
- Max Inf Disintegral Difference = - (Inf Disintegral Results - Min Inf Disintegral Efficiency)
- Inf Disintegral Over-Unity = ((Max Inf Disintegral Results - Min Inf Disintegral Results) / (Max Inf Disintegral Eff)) - Inf Disintegral Diff X 100 (%)
- Proportion of Smaller Unit of Disintegral = 1X Infinity

- Inf Efficiency - Inf Difference (= Anti-Super-Disintegral, same as Super-Negative Disintegral)

- Min Neg Inf Disintegral Results = -[(Max Inf Disintegral Eff / 2) - Inf Disintegral Diff]
- Max Neg Inf Disintegral Results = -[Min Inf Disintegral Eff - Inf Disintegral Diff]
- Min Neg Inf Disintegral Eff = -[Inf Disintegral Results + Inf Disintegral Diff]
- Max Neg Inf Disintegral Eff = -[(Min Inf Disintegral Results + Inf Disintegral Diff) X 2]
- Min Neg Inf Disintegral Difference = (Inf Disintegral Results - Max Inf Disintegral Efficiency) This has no minus sign in the super-negative disintegral
- Max Neg Inf Disintegral Difference = (Inf Disintegral Results - Min Inf Disintegral Efficiency) This has no minus sign in the super-negative disintegral
- Neg Inf Disintegral Over-Unity = -[((Max Inf Disintegral Results - Min Inf Disintegral Results) / (Max Inf Disintegral Eff)) - Inf Disintegral Diff] X 100 (%)
- Proportion of Smaller Unit of Neg Inf Disintegral = 1X Infinity

THE EQUATIONS AND SOLUTIONS TO UNSOLVED PROBLEMS

Inf Difference - Inf Efficiency (= Super-Antitheory)

- Min Inf Results = Inf Diff - (Max Inf Eff / 2 Infinity)
- Max Inf Results = Inf Diff - Min Inf Eff
- Min Inf Eff = Inf Diff - Inf Results
- Max Inf Eff = (Inf Diff - Min Inf Results) X 2 Infinity
- Min Inf Difference = Inf Results + Max Inf Efficiency
- Max Inf Difference = Inf Results + Min Inf Efficiency
- Super-Anti-Unity = ((Max Inf Results - Min Inf Results) / -(Max Inf Eff)) + Inf Diff X 100 (%)
- Proportion of Smaller Unit = 1X Infinity

Inf Efficiency + Inf Difference (= Super-TOE)

- Min Inf Results = (Max Inf Eff / 2) + Inf Diff
- Max Inf Results = Min Inf Eff + Inf Diff
- Min Inf Eff = Inf Results - Inf Diff
- Max Inf Eff = (Min Inf Results - Inf Diff) X 2 Inf
- Min Inf Difference = Inf Results - Max Inf Efficiency
- Max Inf Difference = Inf Results - Min Inf Efficiency
- Over-Unity = ((Max Inf Results - Min Inf Results) / (Max Inf Eff)) + Inf Diff X 100 (%)
- Proportion of Smaller Unit = 1X Infinity

LOOP 11

Inf Difference − Inf Efficiency (= Anti-Super Disintegral / Super-Antitheory)

- Min Inf Results = Inf Diff - (Max Inf Eff / 2 Infinity)
- Max Inf Results = Inf Diff - Min Inf Eff
- Min Inf Eff = Inf Diff - Inf Results
- Max Inf Eff = (Inf Diff - Min Inf Results) X 2 Infinity
- Min Inf Difference = Inf Results + Max Inf Efficiency
- Max Inf Difference = Inf Results + Min Inf Efficiency
- Super-Anti-Unity = ((Max Inf Results - Min Inf Results) / -(Max Inf Eff)) + Inf Diff X 100 (%)
- Proportion of Smaller Unit = 1X Infinity

Inf Efficiency + Inf Difference (= Super-TOE)

- Min Inf Results = (Max Inf Eff / 2) + Inf Diff
- Max Inf Results = Min Inf Eff + Inf Diff
- Min Inf Eff = Inf Results - Inf Diff
- Max Inf Eff = (Min Inf Results - Inf Diff) X 2 Inf
- Min Inf Difference = Inf Results - Max Inf Efficiency
- Max Inf Difference = Inf Results - Min Inf Efficiency
- Over-Unity = ((Max Inf Results - Min Inf Results) / (Max Inf Eff)) + Inf Diff X 100 (%)

THE EQUATIONS AND SOLUTIONS TO UNSOLVED PROBLEMS

- Proportion of Smaller Unit = 1X Infinity

-Inf Difference + Inf Efficiency (= Antisuper / Super-Disintegral)

- Min Inf Disintegral Results = (Max Inf Disintegral Eff / 2) - Inf Disintegral Diff
- Max Inf Disintegral Results = Min Inf Disintegral Eff - Inf Disintegral Diff
- Min Inf Disintegral Eff = Inf Disintegral Results + Inf Disintegral Diff
- Max Inf Disintegral Eff = (Min Inf Disintegral Results + Inf Disintegral Diff) X 2
- Min Inf Disintegral Difference = - (Inf Disintegral Results - Max Inf Disintegral Efficiency)
- Max Inf Disintegral Difference = - (Inf Disintegral Results - Min Inf Disintegral Efficiency)
- Inf Disintegral Over-Unity = ((Max Inf Disintegral Results - Min Inf Disintegral Results) / (Max Inf Disintegral Eff)) - Inf Disintegral Diff X 100 (%)
- Proportion of Smaller Unit of Disintegral = 1X Infinity

-Inf Efficiency - Inf Difference (= Super-Negative Disintegral)

- Min Neg Inf Disintegral Results = - [(Max Inf Disintegral Eff / 2) - Inf Disintegral Diff]
- Max Neg Inf Disintegral Results = - [Min Inf Disintegral Eff - Inf Disintegral Diff]

- Min Neg Inf Disintegral Eff = -[Inf Disintegral Results + Inf Disintegral Diff]
- Max Neg Inf Disintegral Eff = -[(Min Inf Disintegral Results + Inf Disintegral Diff) X 2]
- Min Neg Inf Disintegral Difference = (Inf Disintegral Results - Max Inf Disintegral Efficiency) This has no minus sign in the super-negative disintegral
- Max Neg Inf Disintegral Difference = (Inf Disintegral Results - Min Inf Disintegral Efficiency) This has no minus sign in the super-negative disintegral
- Neg Inf Disintegral Over-Unity = -[((Max Inf Disintegral Results - Min Inf Disintegral Results) / (Max Inf Disintegral Eff)) - Inf Disintegral Diff] X 100 (%)
- Proportion of Smaller Unit of Neg Inf Disintegral = 1X Infinity

THE EQUATIONS AND SOLUTIONS TO UNSOLVED PROBLEMS

ORIGINAL EQUATIONS SLOTTED INTO EACH MAIN SECTION:

Min Results = (Max Eff / 2) + Diff

Max Results = Min Eff + Diff

Min Eff = Results - Diff

Max Eff = (Min Results - Diff) X 2

Min Difference = Results - Max Efficiency

Max Difference = Results - Min Efficiency

Over-Unity = ((Max Results - Min Results) / (Max Eff)) + Diff X 100 (%)

Proportion of Smaller Unit = 1X

...

Nathan Coppedge

,,,

THE EQUATIONS AND SOLUTIONS TO UNSOLVED PROBLEMS

SOLUTIONS TO LONGSTANDING PROBLEMS

Nathan Coppedge

,,,

SOLUTIONS TO UNSOLVED PROBLEMS IN BIOLOGY

ALL PROBLEMS SOLVED IN <24 HOURS— PROBLEMS TAKEN FROM: <u>List of unsolved problems in biology - Wikipedia</u>

PROBLEM:

<u>Alkaloids</u>

The function of these substances in living organisms which produce them is not known

[1]

SOLUTION: Not synthesizing a common element of metabolism

PROBLEM:

<u>Arthropod head problem</u>

A long-standing zoological dispute concerning the segmental composition of the heads of the various arthropod groups, and how they are evolutionarily related to each other.

Segmentation problem as such.

SOLUTION: Lack of segments was broken up.

PROBLEM:

Basking shark

Only the right ovary in female basking sharks appears to function, the reason is unknown.

Shark whisker attraction null genesis.

SOLUTION: No shark without whiskers reproduces with the left ovary of the basking shark: basking sharks are inbred at an early stage.

PROBLEM:

Biological aging

There are a number of hypotheses why senescence occurs including those that it is programmed by gene expression changes and that it is the accumulative damage of biological processes.

Death without environmental factors.

SOLUTION: Life depends on environment.

THE EQUATIONS AND SOLUTIONS TO UNSOLVED PROBLEMS

PROBLEM:

<u>Blue whale</u>

There is not much data on the sexuality of the biggest animal ever.

Biggest whale unknown mating.

SOLUTION: Place without much kelp that they know and they know they have not mated before.

PROBLEM:

<u>[2]</u>

<u>Botany/plants</u>

What is the exact evolutionary history of <u>flowers</u>, called <u>Darwin's abominable mystery</u>?

Seedlike diaspora

SOLUTION: Seeds were destroyed starting in one place we assume.

PROBLEM:

Butterfly migration

How do the descendants of monarch butterfly all over Canada and the US eventually, after migrating for several generations, manage to return to a few relatively small overwintering spots?

Rare, highly specific adaptation to weather and climate.

SOLUTION: Overbearing genetic selection, limited change in circumstances, or successful adaptation to a variety of weather events. Geniuses at weather who died a lot. Doesn't make it easy, but a large population helps. They may eat what others don't eat or can't eat or don't survive on, or they may undergo multiple mating cycles while migrating.

THE EQUATIONS AND SOLUTIONS TO UNSOLVED PROBLEMS

PROBLEM:

Cambrian explosion

What is the cause of the apparent rapid diversification of multicellular animal life around the beginning of the Cambrian, resulting in the emergence of almost all modern animal phyla?

Unusually efficient use of energy.

SOLUTION: Normally we are not efficient or don't use energy by the standards of the Cambrian.

PROBLEM:

Cell size

How do cells determine what size to grow to before dividing?

Certain growth.

SOLUTION: Uncertainty concerning limitation. Natural attenuation. Concern. Rationality.

Nathan Coppedge

PROBLEM:

Cognition and decisions

How and where does the brain evaluate reward value and effort (cost) to modulate behavior? How does previous experience alter perception and behavior? What are the genetic and environmental contributions to brain function?

Higher functioning.

SOLUTION: Instinct. Bootstrapping from lower cognitive functions and environmental clues. No miracle. Highly inconsistent. Imitation. Ideas.

PROBLEM:

Computational neuroscience

How important is the precise timing of action potentials for information processing in the neocortex? Is there a canonical computation performed by cortical columns? How is information in the brain processed by the collective dynamics of large neuronal circuits? What level of simplification is suitable for a description of information processing in the brain? What is the neural code?

Simply complex is ideal.

SOLUTION: Complexly efficient is not ideal.

PROBLEM:

Computational theory of mind

What are the limits of understanding thinking as a form of computing?

Specific thinking tool.

SOLUTION: Thinking is a general tool of it's own general type.

PROBLEM:

Consciousness

What is the brain basis of subjective experience, cognition, wakefulness, alertness, arousal, and attention? Is there a "hard problem of consciousness"? If so, how is it solved? What, if any, is the function of consciousness?

[3]

[4]

Subjective etc awareness.

SOLUTION: Objective realization of limits, for example entities, concepts, missions, historical dialectics.

THE EQUATIONS AND SOLUTIONS TO UNSOLVED PROBLEMS

PROBLEM:

Diseases

What are the neural bases (causes) of mental diseases like psychotic disorders (e.g. mania, schizophrenia), Parkinson's disease, Alzheimer's disease, or addiction? Is it possible to recover loss of sensory or motor function?

Not-typical mental situation.

SOLUTION: Specific situations, rationality, choosing outcomes. Physical limitations and obligations, authentic commitments, loaded plate, unobviousness, thoughtfulness, level of apparent commitment.

PROBLEM:

Evolution of sex

What selective advantages drove the development of sexual reproduction, and how did it develop?

Question of motivation for stimulus.

SOLUTION: Without stimulus, there was less motivation. Motivation was advantageous.

PROBLEM:

Extraterrestrial life

Might life which does not originate from planet Earth also have developed on other planets? Might this life be intelligent?

Similarity of alien life.

SOLUTION: Alien life is different, and might not benefit humans.

PROBLEM:

Facetotecta

The adult form has never been encountered in the water, and it remains a mystery to what it grows into.

Small slug (parasite).

SOLUTION:

Preys on or is eaten by bony fish.

THE EQUATIONS AND SOLUTIONS TO UNSOLVED PROBLEMS

PROBLEM:

Free will

Particularly the neuroscience of free will

Freedom in human biology as explained by neuroscience.

SOLUTION: It is determined by alien non-biological factors which we cannot explain neurochemically. In other words, it is situational, it depends on individualism, and there is no getting rid of it. It is important.

PROBLEM:

Gall wasp

It is largely unknown how these insects induce gall formation in plants; chemical, mechanical, and viral triggers have been discussed.

Old fly layered carapace.

SOLUTION: Newly settled wax example. In other words, stolen flesh and other materials. (Carbon content? Bark?)

PROBLEM:

General anesthetic

What is the mechanism by which it works?

Wave of relief, no sensation.

SOLUTION:

Cuts to the quick nerves.

PROBLEM:

Glycogen body

The function of this structure is not known.

Structure that develops especially in heavy and light dinosaurs where much energy is present.

SOLUTION: Creates sustained effect involving heaviness in light animals and lightness in heavy animals, in other words slight balast effect without much cost of energy.

THE EQUATIONS AND SOLUTIONS TO UNSOLVED PROBLEMS

PROBLEM:

<u>Golgi apparatus</u>

In <u>cell theory</u>, what is the exact <u>transport mechanism</u> by which proteins travel through the <u>Golgi apparatus</u>?

Protein modulator found in females.

SOLUTION: modulates proteins related to males.

PROBLEM:

<u>Homing</u>

A satisfactory explanation for the neurobiological mechanisms that allow homing, has yet to be found.

Unfamiliar smell.

SOLUTION: Familiar smelling.

PROBLEM:

<u>Homosexuality</u>

What is the cause of homosexuality, especially in the human species?

Explaining loving men.

SOLUTION: Not wanting to explain.

PROBLEM:

<u>Korarchaeota</u>

Their metabolic processes are so far unclear.

Low-energy offers no explaining.

SOLUTION: High energy already explains.

THE EQUATIONS AND SOLUTIONS TO UNSOLVED PROBLEMS

PROBLEM:

<u>Language</u>

How is it implemented neurally? What is the basis of <u>semantic</u> <u>meaning</u>?

Hard things seem too easy.

SOLUTION: Easy things are hard-working.

PROBLEM:

<u>Latitudinal diversity gradient</u>

Why does biodiversity increase when going from the poles towards the equator?

Why doesn't heat mean spilled blood?

SOLUTION: Cold means dead. Less dead more life. Coldness runs against life-blood. Heat is the life-blood. More blood more life. Not against nature, part of nature.

PROBLEM:

Learning and memory

Where do our memories get stored and how are they retrieved again? How can learning be improved? What is the difference between explicit and implicit memories? What molecule is responsible for synaptic tagging?

Lazy memory solution.

SOLUTION: Active memory (high-energy use, such as food consumption and stimulant behavior), T1. People originally slept a lot before stimulants.

PROBLEM:

Loricifera

There are at least 100 species of this phylum (many undescribed), but none of them is known to be present in the fossil record.

Old species uninvestigated.

SOLUTION: New investigation.

THE EQUATIONS AND SOLUTIONS TO UNSOLVED PROBLEMS

PROBLEM:

Movement

How can we move so controllably, even though the motor nerve impulses seem haphazard and unpredictable?

Real movement, or movement in reality.

SOLUTION: Virtually timeless or timeless virtue.

PROBLEM:

Development and evolution of brain

How and why did the brain evolve? What are the molecular determinants of individual brain development?

Directive function of the brain.

SOLUTION: Indirect function evolved from everything else.

PROBLEM:

Neural computation

What are all the different types of <u>neuron</u> and what do they do in the human brain?

Directive elements.

SOLUTION: Contingent on being out of their element. Derivative. Virtual. 'Expert', Palliative. Expiative. Similar purpose may have different functions and levels of reality. While reality varies types will be relative.

PROBLEM:

<u>Neuroplasticity</u>

How <u>plastic</u> is the mature brain?

Quanta of intelligence.

SOLUTION: Quality of challenge. Brains either have problems or they don't, so you might say it is just concerning standards and intelligent standards, which come in many varieties.

THE EQUATIONS AND SOLUTIONS TO UNSOLVED PROBLEMS

PROBLEM:

<u>Noogenesis</u> - the <u>emergence</u> and <u>evolution</u> of <u>intelligence</u>

What are the laws and mechanisms - of new <u>idea</u> emergence (<u>insight</u>, creativity synthesis, <u>intuition</u>, <u>decision-making</u>, <u>eureka</u>); development (evolution) of an individual <u>mind</u> in the <u>ontogenesis</u>, etc.?

SOLUTION: Original intelligence, unoriginal mind, original mind, original thought. Essentially, ideas develop as idea-culture not necessarily completely distinct from other culture (sex, entertainment, media, consumerism), however receptiveness to new ideas depends on biology (chemical intake, chemical synthesis, exaggeration, critical faculties, frontal and cerebral brain development, secondary brain development, and repeated exposure, low stress).

PROBLEM:

Origin of life

Exactly how and when did life on Earth originate? Which, if any, of the many hypotheses is correct?

Easy beginning. Then they died.

SOLUTION: Life was hard. It ended many times. It depends on your definition of survive. If you mean big we're dead, if you mean tough we're dead. If you mean smart we've been alive for maybe 12,000 to 120,000 years. But that sounds relative. Also, we could use better history recording still at this point. We might not know without a record of the smart things that happened and a sense of whether they were smart or not.

THE EQUATIONS AND SOLUTIONS TO UNSOLVED PROBLEMS

PROBLEM:

Paradox of the plankton

The high diversity of phytoplankton seems to violate the competitive exclusion principle.

Competitive production against plankton.

SOLUTION: Un-competitive production of plankton.

PROBLEM:

Perception

How does the brain transfer sensory information into coherent, private percepts? What are the rules by which perception is organized? What are the features/objects that constitute our perceptual experience of internal and external events? How are the senses integrated? What is the relationship between subjective experience and the physical world?

High-level integration

SOLUTION: Entry-level problems.

PROBLEM:

<u>Protein folding</u>

What is the folding code? What is the folding mechanism? Can we predict the native structure of a protein from its amino acid sequence?

Amino acid formation blueprint.

SOLUTION: Metabolic shutdown depends on externals. Open system. Stimulus. Food intake. Radically-contingent. Brain patterns.

PROBLEM:

<u>Rotifer</u>

What is the function of the retrocerebral organ?

Nerve sack.

SOLUTION: Stupid gland. This may serve a generic function related to brain function, analogous to grease, stress, or de-stress.

THE EQUATIONS AND SOLUTIONS TO UNSOLVED PROBLEMS

PROBLEM:

<u>Sleep</u>

What is the biological function of sleep? Why do we <u>dream</u>? What are the underlying brain mechanisms? What is its relation to <u>anesthesia</u>?

Why sleep?

SOLUTION: Not why sleep answer. In other words, sleep is for problem-solving, to help us think, perhaps influenced by the concept of drama and stimulation.

…

SOME SOLUTIONS TO SIMILAR PROBLEMS (LIFE SCIENCES) HERE:

To be overly general, some of the difficult issues now or recently are / were:

- How to find new drugs (use available energy input only then hack the energy type).
- How to cure / treat heritable disorders (give gene therapy to successful people on the basis of the technicality of their professed success).
- Questions of how genes work such as epigenetics (variable genius strategies meeting their doom: respond).
- How to build cyborgs (comfort, design, signs, protocol: all four for cyborgs).
- How to cure aging (keep it natural, best herbs such as Jiaogulan, allow strategy).
- How to solve ecological problems, generally involving technology (build a sustainable genius environment).
- How to preserve humans in a spaceship while maintaining stasis (a lot of heat, climate conditioned, with slow takeoff).
- How to improve human intelligence genetically (smart genes that fit

unique footprint and can be delivered by parents).
- How to improve human intelligence with drugs (confirm deep biology, provide needed resources).
- How to allow humans to sustain pleasure (vegetarianism, supportive environment).
- How to improve human libido (healthy options, safe environments).
- How to prevent disease from spreading (reducing vectors, improving sanitation, taking precautions).
- How to cure pandemics (eliminating cause and innoculating).
- How to cure the common cold (Jiaogulan and brainiacs, limited contact).
    —Nathan Coppedge's answer to Where can I find a list of the most difficult concepts/theories in the life sciences?

—SOLUTIONS FROM:

(Sept 27, 2018)

[END OF PROBLEMS IN BIOLOGY]

Nathan Coppedge

,,,

SOLUTIONS TO UNSOLVED PROBLEMS IN CHEMISTRY

(Problems via Wikipedia: List of unsolved problems in chemistry - Wikipedia).

Physical Chemistry

PROBLEM: What are the electronic structures of high-temperature superconductors at various points on their phase diagrams?[1]

SOLUTION: Solution involving the value of relatively low-temperature electrons and magnets (possibly super-solids?).

PROBLEM: Can the transition temperature of high-temperature superconductors be brought up to room temperature?[1]

SOLUTION: Pidgeon: Consistent if low wavelength magnets, electrons, (supersolids?).

PROBLEM: Is *Feynmanium* the last chemical element that can physically exist? That is, what are the chemical consequences of having an element with an atomic number above 137, whose 1s electrons must travel faster than the speed of light?[1][2]

SOLUTION:

Beyond matter is hyper-matter.

PROBLEM: Is Neutronium-4 possible? Stable reactive dichotomy?

SOLUTION: It is stable, but it reacts, perhaps no dichotomy.

PROBLEM: How can electromagnetic energy (photons) be efficiently converted to chemical energy? For instance, can water be efficiently split to hydrogen and oxygen using solar energy?

SOLUTION: Valuable conductor involving non-transcient bases.

THE EQUATIONS AND SOLUTIONS TO UNSOLVED PROBLEMS

Organic chemistry

PROBLEM: Is an *abiologic origin of chirality* as is found in (2*R*)-2,3-dihydroxypropanal (D-glyceraldehyde), and also in amino acids, sugars, etc., possible?[5] Chirality content question regarding structural energy.

SOLUTION: The answer is that what it does not contain is not regarded as non-structural energy, and is not reversed.

PROBLEM: Why are accelerated kinetics observed for some *organic reactions at the water-organic interface*?[6][*non-primary source needed*] Accelerated water reaction organic.

SOLUTION: When it does not react to heat the organic material begins to die, (which can have kinetic results).

PROBLEM: What is the *origin of the alpha effect*, that is, that nucleophiles with an electronegative atom with lone pairs adjacent to the nucleophilic center are particularly reactive?[*citation needed*] Dual repeated weak but reactive problem.

SOLUTION: Concordant effect with grouped dualities en homogeneous masse, group behavior.

THE EQUATIONS AND SOLUTIONS TO UNSOLVED PROBLEMS

Biochemistry problems

PROBLEM: <u>Enzyme kinetics</u>: Why do some enzymes exhibit faster-than-diffusion kinetics?[7]

SOLUTION: Slower metabolic absorption, or non-physical, or hyper-matter.

PROBLEM: <u>Protein folding</u> problem: Is it possible to predict the secondary, tertiary and quaternary structure of a <u>polypeptide</u> sequence based solely on the sequence and environmental information? Inverse protein-folding problem: Is it possible to design a polypeptide sequence which will adopt a given structure under certain environmental conditions?[5][8] This has been achieved for several small globular proteins in recent years.[9] Peptide layer predictions problem.

SOLUTION: One layer cannot be predicted (alone).

PROBLEM: RNA folding problem: Is it possible to accurately predict the secondary, tertiary and quaternary structure of a poly-ribonucleic acid sequence based on its sequence and environment? Atypical expected responses from classic structure?

SOLUTION: Typical adaptive responses to external environment.

PROBLEM: What are the chemical origins of life? How did non-living chemical compounds generate self-replicating, complex life forms?

SOLUTION: Non-synthetic 4-d foundation involving evolution. Life arose from non-synthetic foundations which (sometimes) lacked functions.

PROBLEM: Protein design: Is it possible to design highly active enzymes *de novo* for any desired reaction?[10] Are peptides reactive-generat?

SOLUTION: Requires more than one generation process.

THE EQUATIONS AND SOLUTIONS TO UNSOLVED PROBLEMS

PROBLEM: <u>Biosynthesis</u>: Can desired molecules, natural products or otherwise, be produced in high yield through biosynthetic pathway manipulation?[11] Bio-generation limits problem or not?

SOLUTION: Stopping accumulation of necretic material may create unlimited growth and may be problemetic.

[END OF PROBLEMS IN CHEMISTRY]

Nathan Coppedge

,,,

SOLUTIONS TO UNSOLVED PROBLEMS, COMPUTER SCIENCE

(Problems via Wikipedia: List of unsolved problems in computer science - Wikipedia)

Computational Complexity:

P versus NP problem Compile program faster than time to solve problem program solves?

Decompile random problems is slower because perfect problems solve random problems through problematics. Arrange a perfect problem to solve all random problems. Smaller problems may be more difficult, like crossing a stream by walking on a twig. Previous more detailed answer on this: Attempt to Solve the P Versus NP Problem

What is the relationship between BQP and NP?

The quantum computer seems to give the right answer the wrong way, to the extent that it's the right answer. On the other hand, it will give the wrong answer wrong with 100% probability if it is right. However, there may be some metaphysical concerns. If you really have a problem it will solve it particularly if you don't know what you're doing, but if you have a perfect problem it will never knowingly be solved by a quantum computer unless you are looking for an imperfect solution. Recording quantum logic linguistically provides an alternative (see P vs. NP above).

NC = P problem
Chain of the organization ^ organization must be synchronized (like a metaphor).

NP = co-NP problem
Some problems are complete, unless they are defined as incomplete.

THE EQUATIONS AND SOLUTIONS TO UNSOLVED PROBLEMS

P = BPP problem
BPP (random) is complete if it happens that way. Random is not random if completeness is / becomes the rule, for example evolutionarily.

P = PSPACE problem
Examples will arise, however their larger set will rarely if ever be complete.

L = NL problem
Conjectural completeness manifest, def y = def x (y).

PH = PSPACE problem
For class N (infinity) complexity is absolutely trivial.

L = P problem
Effect of a logarithm must be equivalent, effect can be simulated.

L = RL problem
Solution: Identity as the solution.

Unique games conjecture
Defined cases do not require uniqueness if P permits a larger case than that non-uniqueness.

Is the exponential time hypothesis true? Is the strong exponential time hypothesis (SETH) true?

No, expenential efficiency is possible because it is a concept and I have thought of it. See under P vs. NP.

Do one-way functions exist? Is public-key cryptography possible?

You simply need a dysfunction that decomposes. Reliability is the problem, rather than penetrability. Perhaps making better inferences such as location tracking, audio recording, behavior detection, and perhaps genetics. Also, encryption is sometimes overrated.

THE EQUATIONS AND SOLUTIONS TO UNSOLVED PROBLEMS

Polynomial versus non-polynomial time for specific algorithmic problems

Main article: NP-intermediate

Can integer factorization be done in polynomial time on a classical (non-quantum) computer?

Some possible clues as far as irrational numbers: It turns out if I take (((10e) (pi - 1) *X) / 2) this produces some differences in sequence near the decimal point, but if I then * (e / pi) the result is an interesting number: 25.18523247... As it turns out, the first two digits are 5 * 5, which I call an irrational factorization of 1 * 10. The second two digits are equal to 2 * 9, the fifth and sixth digits are a straight irrational factorization, the seventh and eight digits are 3 * 8. Get this, the ninth and tenth digits are 4 * 7.

Can integer factorization be considered to be NP-complete?

No, because there will be some reference frame from which they are P-incomplete.

Can <u>clustered planar drawings</u> be found in polynomial time?

Yes, individual axial relationships can be found, and expressed as a sum.

Can the <u>discrete logarithm</u> be computed in polynomial time?

Something similar to y / x^2 can be used to find the slope of a logarithm.

Can the <u>graph isomorphism problem</u> be solved in polynomial time?

If a characterization and a limit can be reached so as to express the whole graph, the limit can be characterized.

Can <u>leaf powers</u> and k-leaf powers be recognized in polynomial time?

If the largest group (total group) is first recognized, an additional step can be used to collect axes which may be expressed as a sum.

THE EQUATIONS AND SOLUTIONS TO UNSOLVED PROBLEMS

Can parity games be solved in polynomial time?

It's restricted by the modification of the range of the range, reapplied and or modified N number of times. Other than this formula it may have other required determinations or not, for example, concepts. As an open-ended game it has open-ended results of the kind described. As a free will game the game is potentially open, except for range, range of range, and limits on will. Etc for other conditions which may be different. Variation on the modification of the range of the range, reapplied and or modified N number of times seems to describe the whole set. If one answer is incorrect, it may easily be the limit of another answer. In this way the formula is correct. Alternately, another option is that the game is never really played, and movements are not compulsory, so between limits and no compulsoriness the game is undecidable if any moves are available (unless there is no range at all), as moves are optional or become optional.

Can the <u>rotation distance</u> between two <u>binary trees</u> be computed in polynomial time?

Cascade problem indicates this is NP-complete unless the Cascade problem has a more efficient solution. "Cantor assumes the sets are not value-ordered." —<u>The Ultimate Critique of Mathematics</u>

Can graphs of bounded <u>clique-width</u> be recognized in polynomial time?[1]

To the extent that they can be recognized at one level, with enough consistency a token theory is highly probable. One approach is to make this problem subsistent on other problems so outside data can be used to solve the problem more simply.

Can one find a <u>simple closed quasigeodesic</u>on a convex polyhedron in polynomial time?[2]

I suppose you could take the Fibonacci number of a side of a tetrahedron filling a sphere and modify the slope such as to occupy the entire triangulated surface of the sphere, then use avoidance to create three or more connected shapes with a probability of $1/(x+3)$ coverage.

THE EQUATIONS AND SOLUTIONS TO UNSOLVED PROBLEMS

Can a <u>simultaneous embedding</u> with fixed edges for two given graphs be found in polynomial time?[3]

If they are the same shape (same data), if shapes may be clusters of shapes fastest formula may be x not= x2, y not=y2. This can be done in sectors unless the resolution is infinite. If different shapes, it is indeterminate because we don't know the level of resolution.

Other algorithmic problems

What is the fastest <u>algorithm for multiplication</u> of two n-digit numbers?

The fastest formula assuming difficulty is to be really lucky or to have the data stored in a high-speed database or as a logarithm.

What is the fastest <u>algorithm for matrix multiplication</u>?
Add the column labels. Multiply the column total by each row label to yield the total for each row. Divide the total for each row by the column labels to yield the number in each box.

Can the <u>Schwartz–Zippel lemma</u> for <u>polynomial identity testing</u> be <u>derandomized</u>?

Likely meaningless.

THE EQUATIONS AND SOLUTIONS TO UNSOLVED PROBLEMS

Can a <u>depth-first search tree</u> be constructed in <u>NC</u>?

Using ideas or identities, such a graph can be represented symbolically with such phenomenal relationships, on the assumption it is infinite. That is, infinity can be represented and inflected symbolically.

Does <u>linear programming</u> admit a <u>strongly polynomial</u>-time algorithm? This is problem #9 in <u>Smale's list</u> of problems.

Very similar to P vs. NP. This might be a question of aesthetics.

What is the lower bound on the complexity of <u>fast Fourier transform</u> algorithms? Can they be $o(N \log N)$?

Might require modifying Fourier for use of slope.

The dynamic optimality conjecture: do splay trees have a bounded competitive ratio?

Splay trees might obviously be limited if P = NP.

Can we compute the edit distance between two strings of length *n* in strongly sub-quadratic time, i.e., in time $O(n2-\epsilon)$ for some $\epsilon>0$?

Sounds like a hardware problem.

Is there a k-competitive online algorithm for the k-server problem?

Appears to be resolved with 2n +1 and $\log^2 k \log^3 n$.

THE EQUATIONS AND SOLUTIONS TO UNSOLVED PROBLEMS

Can X + Y sorting be done in better than *o* (*n*2 log *n*) time?

Another case where a simple problem may be harder to solve. Some formulas may make much tougher data easy to find. In sny case, if the method is that efficient the question is merely exponential efficiency or not, and I am not sure of the technical specifics.

How many queries are required for envy-free cake-cutting?

Likely 2n, where n is the number of pieces. If you get it wrong you try again.

What is the lowest possible average-case time complexity of Shellsort with a deterministic, fixed gap sequence?

It is completed in average height * width * random (X) for AI capability (+1 for filling the whole screen automatically if you lose).

79

Natural Language Processing algorithms

Is there any perfect <u>syllabification</u> algorithm in the English language?

They just have to be assigned based on standards, because standards can be very high, but usually are not as complex as the very most complex. I can imagine a version where letters are placed on top of one another almost like Chinese, but that is not what most people want.

Is there any perfect <u>stemming</u> algorithm in the English language?

Stemming is questionable as there may be better organizational systems like polar opposites, noun & property, popularity, system.

THE EQUATIONS AND SOLUTIONS TO UNSOLVED PROBLEMS

Is there any perfect <u>POS tagging</u> algorithm in the English language?

An ideal simplified form is 1. Nouns including proper names, 2. Qualities (Adjectives and verbs), 3. Booleans (because, with, but, and, is, as is, just as, when, so, and as such…), 4. Heuristics (compounds forming deeper meaning sometimes using perfected language formulas).

Programming language theory

Main article: Programming language theory

POPLmark
Programmable Heuristics.

Barendregt–Geuvers–Klop conjecture
By 'strongly normalizing' what might be implied is type identity consistency, but this ostensibly controvenes the possibility of multi-theoretics, suggesting they are not opening themselves to critical analysis.

Other problems

Aanderaa–Karp–Rosenberg conjecture
Vectors or at least directions guarantee that graphs require investigation unless perhaps as my brother the programmer says, properties are localized somehow. Or I might add invisibly structured or preconceived.

Generalized star height problem
(Interesting abstract problem). Clearly Kleene Stars have something to do with coherent depth. This form of 'height' or obscurity means very little without coherence.

Separating words problem
Possibility of hypercomputation

[END OF PROBLEMS IN COMPUTER SCIENCE]

Nathan Coppedge

SOLUTIONS TO UNSOLVED PROBLEMS IN ENVIRONMENTALISM

In progress, problems take from: List of environmental issues - Wikipedia

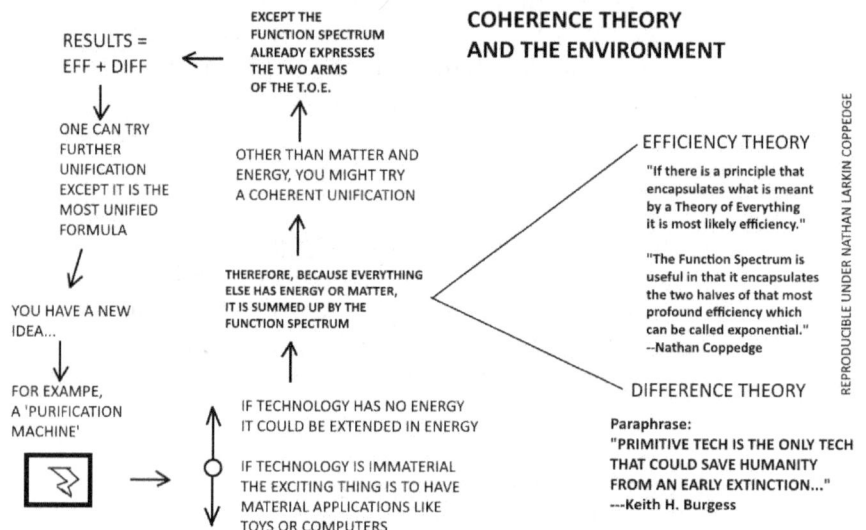

Human overpopulation — Biocapacity.

Solution: Food supply is larger than thought. Create economic resources to pay for higher food production. Use mineral fertilizers, produce foods like mushrooms that can be grown indoors in large quantities. Increase urban food production. Cut down on demand. Provide chemical alternatives to meals. Try replacing some foods with stimulants. Reduce calorie requirements by reducing athleticism. Produce foods under a centralized program to improve bulk purchasing efficiency. Improve inherent quality and cheapness of grains and fruit. Create edible weeds. Increase use of a liquid diet which is fulfilling yet tasty.

Principle: Sometimes 'wows' come from wherever wows come from. For example, the PC revolution may have reduced overpopulation.

THE EQUATIONS AND SOLUTIONS TO UNSOLVED PROBLEMS

Someone using a computer might think it deserves a 'wow' to notice that.

<u>Human overpopulation</u> — <u>climate change</u>

Increase use of indoor living. Use solar. Build solid infrastructure that doesn't need many repairs or can be repaired cheaply with minimal loss of life. Attract talent to safe areas to preserve humanity. Use economic and government resources for overall success. Plan brilliantly, dream appropriately.

<u>Human overpopulation</u> — <u>Water scarcity</u>

Get high-end market to buy special water. Use profits from rich to help the poor, possibly even using salinization by boiling sea water. Earth has a lot of water, there ought to be ways to make use of it. Encourage resettlement in areas with greater water supplies. Hook people on high-end products that limit consumption and put a price on the commodity. Give incentives for wealthier people to help redistribute water from mountain springs. Prevent absorption and evaporation of water in arid areas.

<u>Intensive farming</u> —
Solution: Leave it to the farmers. Provide monetary incentives. Work with the best food producers, importing if possible.

<u>Land use</u> — <u>Desertification</u>

Solution: Prevent runaway soil by reducing deforestation and encouraging 'survival species' of plants. Avoid pesticides. Use special kinds of complex or compound soil that stay clumped together.

Land use — Habitat destruction

Encourage major bird species to survive. Support the continuation of sustainable fishing. Create zoos for exotic animals. Support the environmental agencies and organizations, and supply funds for sustainable research that doesn't involve killing animals.

Natural disasters

- Earthquakes

Solution: Buildings suspended on long-distance cables, using land that is relatively immune to earthquakes.

- Tsunamis

Solution: Relocating to higher ground or conducting activity in a covered, reinforced environment.

- Asteroids

Solution: Reduce risk where possible, by choosing a safe planet and gradually destroying high-risk asteroids where feasible.

- Land-damage Hurricanes

Solution: Build buildings to withstand weather and to be immune to flooding, or build further inland.

- Sea-damage Hurricanes

Solution: Avoid risking loss of ships and cargo. Use submarines, or try air and land transport where risk is too much, or create safer ships.

- Tornadoes

Solution: Go underground, fortify. Build strongly or cheaply. Avoid worst areas, or do something with the air pressure to prevent tornadoes.

- Flooding.

In urban areas, use a sewer system and a pump-drainage system to help limit overflow. Pump into a slightly higher enclosed, flat storage area. Build land higher where possible with a pavement covering and significant pylons, earthquake proof if necessary. Create natural flow onto lower land or into the ocean. Build a strong larger framework between buildings to allow building higher, this may require incentives for bulk purchases of land and simultaneous meeting of local needs by reduction of land waste and subsidy programs. Use sandbags and other barriers to prevent flow of water towards urban areas.

- Landslides.

Solution: Seek higher ground, or ground that is less wet. Avoid driving in the rain when there are hills next to the road or large bodies of water nearby. Use an all-terrain, amphibious, floating, or flying vehicle. Avoid worst areas it might not be survivable.

...

Nuclear issues — Nuclear weapons

Safe politics, policing of terrorism, avoiding extremism, economic motives, trade agreements, international treaties / agreements, disarmament, passive role of weapons, failsafes, computerized security calculations focused partly on preserving the planet and unseen consequences, economic warfare against nuclear nations, appropriate philosophical distractions.

Nuclear issues -- Radiation accidents

Containment. over-landing. Change water supply. Rely on pre-manufactured items. Avoid worst areas. Develop anticancer genes. Survival genes. Nutrition with very minimal intake. Buy cheap foreign products. Avoid traveling the area a lot. Get help from animals and experts. Eventually, tolerate mutations or design humans for better survival without mutations.

Ocean trash
Solution: Species will adapt, but we may not like it, so we should create an environment better for the species we prefer. *This may for example involve bigger subsidies for the fishing industry, greater support of genetically-modified fish, and the creation of special environments for species-testing. Also, programs are under way to sweep garbage using big barriers and collect it in certain areas. This effort may be partly successful. To use only biodegradable plastic bags may be important, involving some type of cheap synthesization, I think this is already existing. And, to make sure Starbux uses biodegradable materials (currently it doesn't).*

Water Pollution
To use eco-friendly dish detergent, avoid sending large amounts of paint into the ocean, avoid using nuclear energy which involves radioactive waste water, reduce the dependence on harmful industrial products like WD-40.

...

OTHER ENVIRONMENTAL SOLUTIONS:

Existence is sustainable by whatever qualifications apply to it immortally.

(TENTATIVE) SOLUTIONS TO UNSOLVED PROBLEMS IN HUMANITARIANISM

No wikipedia article currently exists on unsolved problems in humanitarianism.

This writing is currently not complete, although some things are covered by the separate list of solutions on environmentalism.

Abstract / obtuse division prevents problem-solving.

A dichotomy between privilege and neediness prevents social solutions.

Schizophrenia or weakness prevents genuine telepathy.

Practical limits prevent great human intelligence.

--What-prevents-human-beings-from-thinking-together/answer/Nathan-Coppedge

Nathan Coppedge

SOLUTIONS TO UNSOLVED PROBLEMS IN MATHEMATICS

Problems taken from Wikipedia

Solutions to Unsolved Problems

Algebra

PROBLEM: Homological conjectures in commutative algebra

SOLUTION: (1) A dissipated analogue, (2) Pure math, or (3) Simplicity.

Finite lattice representation problem Specific bubble problem.

SOLUTION: Relatively speaking, yes. Or roughly speaking, maybe.

Hilbert's sixteenth problem. Length of incoherent curves.

SOLUTION: They are coherent or straight meaning moduar in length. I suggest modulo 2.

Hilbert's fifteenth problem

SOLUTION: Cyclical morphology (perhaps involving i), d-dimensional, applied to a surface or a trans-finite lattice, or expressed as limit chain functions.

THE EQUATIONS AND SOLUTIONS TO UNSOLVED PROBLEMS

Hadamard conjecture. Interpreted as 1's and 0's which have a greater sum if they fall within a square grid.

SOLUTION: In terms of limit curves, yes. In terms of area, no.

Jacobson's conjecture

SOLUTION: This may require calling a theophrastus, which in this case means calling an open set on the limit function. Then a location can be called, which arbitrarily refers to the 2-d Noetherian ring. The location is like an undefined variable. Using the theophrastus makes a solution more imaginable, it is represented by a figure in the shape of a blank thick column with unmarked angled top and base.

Crouzeix's conjecture

SOLUTION: Of course, if we assume a bounded infinite matrix, because only half the values will not be limits. If the matrix is unbounded it might all be limits, and if it is not infinite, it might all be algebra. The problem beyond this is if its a fuzzy matrix, which might not be Hermetian.

Existence of perfect cuboids and associated cuboid conjectures

SOLUTION: A coherent cube or coherent hypercube maximizes the number of space diagonals in a perfect cuboid. However, it might require conceptualism or relativity. Interestingly, the coherent cube might be expressed as a dimensional limit on a coherent sphere.

Zauner's conjecture: existence of SIC-POVMs in all dimensions

SOLUTION: The quantum-quantum observation problem has to do with absolute quantum states and the condition of non-observability. By pseudo-definition, such states will be absolutely quantum, OR non-observable, either of which may be qualified opposite of the conditionality, or else qualified non-objectively. As some have said, if you study quantum, it's not a problem, but if you don't, it's a big problem. But what we mean by studying in this sense is itself not quantum, or at least not objective, if the aim is objectivity. One solution is quantum must choose between objective and subjective, but as soon as it is physics, it has really chosen the opposite of it's choice, or some larger condition holds on the observations. Unless we define the larger ccnditions, there is some property that has not been observed.

Wild Problem: Classification of pairs of n×nmatrices under simultaneous conjugation and problems containing it such as a lot of classification problems

POSSIBLE SOLUTION: An intuition is it involves straight rows and columns, perhaps selectively. It may be a case of non-exclusive universalism.

Köthe conjecture

For symmetries N and asymmetries S, either there will be symmetry or asymmetry in some degree. If this is true, and the object is a ring, it appears that rings are nil, assuming space. However, secondary symmetries and primary asymmetries could exist which subjectively favor a different nil. Therefore, secondary nils appear to be the exception, raising vocabulary like locally closed under transfinites. There is also a question of negative dimensions or extended nils. Extended nils might render the problem relativistic. Consider a loop asymmetrically wrapped in a spiral around an asymmetrically curved second length, both lengths forming somewhat of a 'C' shape. It seems to me there could be two- or three- or greater-dimensional non-nil centers in this case. Consider that part of the second length which composes the spiral is 3-dimensionally parallel to the curved first member, but the spiral occurs at an asymmetric length along the first member, and is offset inward from the initial ends of the second length. Then multiple parts of the second length might be equidistant to a part of the first length, but asymmetrically.

Birch–Tate conjecture

If the Galois group is the inverse function group, it makes sense that the coherent zeta function applied to the Galois group would be the inverse zeta function.

Serre's conjecture II

SOLUTION: Assuming H1 means hierarchy level 1, "H1(F, G)" is not zero under coherent asymmetric numbers. Or, at least, if the perfect group is incoherent (not absolutely coherent), it is incoherent or non-philosophically-absolute.

Bombieri–Lang conjecture

SOLUTION: The solution states that given the plausability of a coherent case K, the coherence of K scales to the contents of K. However, since the contents may be viewed incoherently, there is no objective cohomology of K which proves K is coherent, because coherence is just an assumption. Nor can we prove K is non-dense without coherence or something similar. Since something similar may make the same formal assumptions as coherent K, the theorem is naive until it is decided if there is a coherent K or something similar.

Farrell–Jones conjecture

Bost conjecture

Rota's basis conjecture

Uniformity conjecture

Kaplansky's conjecture

Kummer–Vandiver conjecture

Serre's multiplicity conjectures

Pierce–Birkhoff conjecture

Eilenberg–Ganea conjecture

Green's conjecture

SOLUTION: For arbitrary curves, a meaningful value might be had by (g - 1/2 g / a length ratio typology) / g, applied recursively as necessary.

THE EQUATIONS AND SOLUTIONS TO UNSOLVED PROBLEMS

Grothendieck–Katz p-curvature conjecture

Sendov's conjecture

Zariski–Lipman conjecture

Algebraic geometry

Abundance conjecture

Bass conjecture

Deligne conjecture

Fröberg conjecture

Fujita conjecture

Hartshorne conjectures

The Jacobian conjecture

Manin conjecture

Nakai conjecture

Resolution of singularities in characteristic p

Standard conjectures on algebraic cycles

Section conjecture

Tate conjecture

Termination of flips

Virasoro conjecture

Zariski multiplicity conjecture

Analysis

The four exponentials conjecture on the transcendence of at least one of four exponentials of combinations of irrationals[13]

Lehmer's conjecture on the Mahler measure of non-cyclotomic polynomials[14]

The Pompeiu problem on the topology of domains for which some nonzero function has integrals that vanish over every congruent copy[15] "Tessellating identity problem"

SOLUTION: "What does not tessellate will not integrate or will have a solution."

Schanuel's conjecture on the transcendence degree of exponentials of linearly independent irrationals[13] "Number theory break it down".

SOLUTION: Graph theory join it up.

Are γ

(the Euler–Mascheroni constant), π + e, π − e, πe, π/e, πe, π√2, ππ, eπ2, ln π, 2e, ee, Catalan's constant, or Khinchin's constant rational, algebraic irrational, or transcendental? What is the irrationality measure of each of these numbers?[16][17][18]

SOLUTION: They are all associated with the universal-universal constant, no surprise. See Sublime Constants, Coppedge.

THE EQUATIONS AND SOLUTIONS TO UNSOLVED PROBLEMS

Vitushkin's conjecture

Combinatorics

Frankl's union-closed sets conjecture: for any family of sets closed under sums there exists an element (of the underlying space) belonging to half or more of the sets[19]

SOLUTION: It is as if these sets are not closed under divisibility. Of course not, they have quantity. If the problem is trans-finites, proportional numbers may be required to help complete number theory. There is no conceptual limitation to creating numbers which fall between infinites and finites, it simply involves infinite ratios.

The lonely runner conjecture: if $k+1$

runners with pairwise distinct speeds run round a track of unit length, will every runner be "lonely" (that is, be at least a distance $1/(k+1)$

from each other runner) at some time?[20]

SOLUTION: Can you artificially set it up? If not, what is the point in believing it is possible? You can certainly create symmetries of sorts, but unless you define the system so that symmetric speeds are not synchronized, there will be no way to solve the problem. If all the speeds are different, they can be radially symmetrical for certain. If the speeds are relative, then there may be ways of creating fixed points or even making all the dots remain in one place. Some trickery is required evidently.

Singmaster's conjecture: is there a finite upper bound on the multiplicities of the entries greater than 1 in Pascal's triangle?[21]

SOLUTION: No, because the hierarchy can repeat in trans-finite dimensions. However, it may depend on specific conjectures in number theory.

Finding a function to model n-step self-avoiding walks.[22]

SOLUTION: As much space as possible or cheating enough. Hard to be more general than that. Complex spatial rules may constrain solutions to arbitrarily complex cheating, thus, this may be the best possible solution for all spatial geometries with all possible cheating. For example, language cheating or cheating using puzzle-solving. Also, alternate answers are likely to be more complex and harder to learn. Restricting it to a narrow case like a chess board has the effect of making the solution less general.

The 1/3–2/3 conjecture: does every finite partially ordered set that is not totally ordered contain two elements x and y such that the probability that x appears before y in a random linear extension is between 1/3 and 2/3?[23]

SOLUTION: No, in 3-d depending on definition, Y could appear before X on average but not in terms of distance. Also, depending on definition, it may be possible to have a backwards-ordered set where Y would appear before X even according to the ordinary order for the reverse direction, and with a different ratio.

THE EQUATIONS AND SOLUTIONS TO UNSOLVED PROBLEMS

Give a combinatorial interpretation of the Kronecker coefficients.[24]

Differential geometry

The filling area conjecture, that a hemisphere has the minimum area among shortcut-free surfaces in Euclidean space whose boundary forms a closed curve of given length[25]

SOLUTION: Limit-curved polygons could contain the same area while not representing the boundary, is my solution. Therefore, it does not have a minimum area under the assumption that one may use equal-summed limit-curved polygons. But they same may be true of any geometric shape 2-d or larger. This may connect with my thought that energy of a dimension equals D - 100%

The Hopf conjectures relating the curvature and Euler characteristic of higher-dimensional Riemannian manifolds[26]

SOLUTION: As far as 4-d and higher isometry, my conclusion is it is from our perspective spaceless and therefore follows a kind of spaceless curve which looks like a straight line from many angles, yet is still curved. I have associated these curves with 'Eridianism', 'Disintegrals', and 'Derigatives' as a kind of non-mathematical geometric entity. A post-mathematical idea related to the true sense of higher dimensions. Another perspective is that it represents 4-d space in the truer sense of 4-d space, analogous to something like 4-d-represented graph theory on 4-category semantics. One way to see this is that the perspective shifts but the perspective is purely objective and isometric.

Nathan Coppedge

The spherical Bernstein's problem, a possible generalization of the original Bernstein's problem

SOLUTION: If ideally physically constructed analogous to mathematics, if we assume spatial continuity and dimensional continuity, the observation of the sacrosanctitude of the dimensional continuum suggests that when it is obvious that dimension four continues to dimension five and that dimension four continues to dimension three as opposed to skipping more than one dimension, then if continuity is analogous to a mobius strip or flat geometry then continuations in eight or nine duplications will have the same properties as four duplications, except with more scattered lines.

Chern's conjecture (affine geometry)

SOLUTION: In pure semantics analogous to a more general case of pure math, any surface can be flat, so most generally it is a semantic concern to ask if that is possible.

Chern's conjecture for hypersurfaces in spheres

SOLUTION: This is similar to the time-travel problem for time-crystals. The surface is infinite if one assumes infinite action on the surface, or infinite this or that like infinite morphization. However, with finite action the surface is never infinite unless the finite action involves infinite ratios.

THE EQUATIONS AND SOLUTIONS TO UNSOLVED PROBLEMS

Yau's conjecture

SOLUTION: By chiral symmetry, this problem appears to be solved provided a continuous surface. In fact, it might even be solved for open figures.

Yau's conjecture on the first eigenvalue

SOLUTION: Seems to depend on whether it is an internal sub-manifold. If the sub-manifold is external, the next manifold may not equal the Eigenvalue of n (1).

Discrete geometry

Solving the happy ending problem for arbitrary n

[27]

Finding matching upper and lower bounds for k-sets and halving lines[28]

The Hadwiger conjecture on covering n-dimensional convex bodies with at most 2nsmaller copies[29]

The Kobon triangle problem on triangles in line arrangements[30]

The McMullen problem on projectively transforming sets of points into convex position[31]

Tripod packing[32]

SOLUTION: Assuming overall length is the constraining factor and a cube is desired, you can rotate them and stack either 2+ the additional segment distance, or 8 + the additional segment distance, depending on whether you start out with a width of twice and a bit more the depth of the tripod or just a bit more than once. The solution is at least related to this, though there may be more multiplication for inside inverses.

Ulam's packing conjecture about the identity of the worst-packing convex solid[33]

Kissing number problem for dimensions other than 1, 2, 3, 4, 8 and 24[34]

How many unit distances can be determined by a set of n points in the Euclidean plane?[35]

The Calluna's Pit problem on geometric probability[36]

Euclidean geometry

Bellman's lost in a forest problem – find the shortest route that is guaranteed to reach the boundary of a given shape, starting at an unknown point of the shape with unknown orientation[37]

Danzer's problem and Conway's dead fly problem – do Danzer sets of bounded density or bounded separation exist?[38]

Dissection into orthoschemes – is it possible for simplices of every dimension?[39]

The einstein problem – does there exist a two-dimensional shape that forms the prototile for an aperiodic tiling, but not for any periodic tiling?[40]

The Erdős–Oler conjecture that when n

is a triangular number, packing $n-1$ circles in an equilateral triangle requires a triangle of the same size as packing n circles[41]

Falconer's conjecture that sets of Hausdorff dimension greater than $d/2$ in R^d

Inscribed square problem – does every Jordan curve have an inscribed square?[43]

The Kakeya conjecture – do n-dimensional sets that contain a unit line segment in every direction necessarily have Hausdorff dimension and Minkowski dimension equal to n?[44]

The Kelvin problem on minimum-surface-area partitions of space into equal-volume cells, and the optimality of the Weaire–Phelan structure as a solution to the Kelvin problem[45]

Lebesgue's universal covering problem on the minimum-area convex shape in the plane that can cover any shape of diameter one[46]

Moser's worm problem – what is the smallest area of a shape that can cover every unit-length curve in the plane?[47]

The moving sofa problem – what is the largest area of a shape that can be maneuvered through a unit-width L-shaped corridor?[48]

Shephard's problem (a.k.a. Dürer's conjecture) – does every convex polyhedron have a net?[49]

The Thomson problem – what is the minimum energy configuration of n

 mutually-repelling particles on a unit sphere?[50]

Uniform 5-polytopes – find and classify the complete set of these shapes[51]

Covering problem of Rado – if the union of finitely many axis-parallel squares has unit area, how small can the largest area covered by a disjoint subset of squares be?[52]

Dynamical systems

Collatz conjecture ($3n + 1$ conjecture)

Lyapunov's second method for stability – For what classes of ODEs, describing dynamical systems, does the Lyapunov's second method formulated in the classical and canonically generalized forms define the necessary and sufficient conditions for the (asymptotical) stability of motion?

Furstenberg conjecture – Is every invariant and ergodic measure for the $\times 2, \times 3$

 action on the circle either Lebesgue or atomic?

Margulis conjecture – Measure classification for diagonalizable actions in higher-rank groups

MLC conjecture – Is the Mandelbrot set locally connected?

Weinstein conjecture – Does a regular compact contact type level set of a Hamiltonian on a symplectic manifold carry at least one peri-

THE EQUATIONS AND SOLUTIONS TO UNSOLVED PROBLEMS

odic orbit of the Hamiltonian flow?

Arnold–Givental conjecture and Arnold conjecture – relating symplectic geometry to Morse theory

Eremenko's conjecture that every component of the escaping set of an entire transcendental function is unbounded

Is every reversible cellular automaton in three or more dimensions locally reversible?[53]

Many problems concerning an outer billiard, for example show that outer billiards relative to almost every convex polygon has unbounded orbits.

Games and puzzles

Sudoku: What is the maximum number of givens for a minimal puzzle?[54] How many puzzles have exactly one solution?[54] How many minimal puzzles have exactly one solution?[54]

Tic-tac-toe variants: Given a width of tic-tac-toe board, what is the smallest dimension such that X is guaranteed a winning strategy?[55]

Graph theory

Paths and cycles in graphsEdit

Barnette's conjecture that every cubic bipartite three-connected planar graph has a Hamiltonian cycle[56]

Chvátal's toughness conjecture, that there is a number t such that every t-tough graph is Hamiltonian[57]

The cycle double cover conjecture that every bridgeless graph has a family of cycles that includes each edge twice[58]

The Erdős–Gyárfás conjecture on cycles with power-of-two lengths in cubic graphs[59]

The linear arboricity conjecture on decomposing graphs into disjoint unions of paths according to their maximum degree[60]

The Lovász conjecture on Hamiltonian paths in symmetric graphs[61]

The Oberwolfach problem on which 2-regular graphs have the property that a complete graph on the same number of vertices can be decomposed into edge-disjoint copies of the given graph.[62]

Graph coloring and labeling

The Erdős–Faber–Lovász conjecture on coloring unions of cliques[63]

The Gyárfás–Sumner conjecture on χ-boundedness of graphs with a forbidden induced tree[64]

The Hadwiger conjecture relating coloring to

THE EQUATIONS AND SOLUTIONS TO UNSOLVED PROBLEMS

clique minors[65]

The Hadwiger–Nelson problem on the chromatic number of unit distance graphs[66]

Hedetniemi's conjecture on the chromatic number of tensor products of graphs[67]

Jaeger's Petersen-coloring conjecture that every bridgeless cubic graph has a cycle-continuous mapping to the Petersen graph[68]

The list coloring conjecture that, for every graph, the list chromatic index equals the chromatic index[69]

The Ringel–Kotzig conjecture on graceful labeling of trees[70]

The total coloring conjecture of Behzad and Vizing that the total chromatic number is at most two plus the maximum degree[71]

Graph drawing

The Albertson conjecture that the crossing number can be lower-bounded by the crossing number of a complete graph with the same chromatic number[72]

The Blankenship–Oporowski conjecture on the book thickness of subdivisions[73]

Conway's thrackle conjecture[74]

Harborth's conjecture that every planar graph can be drawn with integer edge lengths[75]

Negami's conjecture on projective-plane embeddings of graphs with planar covers[76]

The strong Papadimitriou–Ratajczak conjecture that every polyhedral graph has a convex greedy embedding[77]

Turán's brick factory problem – Is there a drawing of any complete bipartite graph with fewer crossings than the number given by Zarankiewicz?[78]

Universal point sets of subquadratic size for planar graphs[79]

THE EQUATIONS AND SOLUTIONS TO UNSOLVED PROBLEMS

Miscellaneous graph theory

Conway's 99-graph problem: does there exist a strongly regular graph with parameters (99,14,1,2)?[80]

The Erdős–Hajnal conjecture on large cliques or independent sets in graphs with a forbidden induced subgraph[81]

The GNRS conjecture on whether minor-closed graph families have $\ell 1$

 embeddings with bounded distortion[82]

The implicit graph conjecture on the existence of implicit representations for slowly-growing hereditary families of graphs[83]

Jørgensen's conjecture that every 6-vertex-connected K6-minor-free graph is an apex graph[84]

Meyniel's conjecture that cop number is $O(n^{--\sqrt{}})$

[85]

Does a Moore graph with girth 5 and degree 57 exist?[86]

What is the largest possible pathwidth of an n-vertex cubic graph?[87]

The reconstruction conjecture and new digraph reconstruction conjecture on whether a graph is uniquely determined by its vertex-deleted subgraphs.[88][89]

The second neighborhood problem: does every oriented graph contain a vertex for which there are at least as many other vertices at distance two as at distance one?[90]

Sumner's conjecture: does every $(2n-2)$

-vertex tournament contain as a subgraph every n

-vertex oriented tree?[91]

Tutte's conjectures that every bridgeless graph has a nowhere-zero 5-flow and every Petersen-minor-free bridgeless graph has a nowhere-zero 4-flow[92]

Vizing's conjecture on the domination number of cartesian products of graphs[93]

THE EQUATIONS AND SOLUTIONS TO UNSOLVED PROBLEMS

Group theory

Is every finitely presented periodic group finite?

The inverse Galois problem: is every finite group the Galois group of a Galois extension of the rationals?

For which positive integers m, n is the free Burnside group B(m,n) finite? In particular, is B(2, 5) finite?

Is every group surjunctive?

Andrews–Curtis conjecture

Herzog–Schönheim conjecture

Does generalized moonshine exist?

Are there an infinite number of Leinster Groups?

Model theory

Vaught's conjecture

The Cherlin–Zilber conjecture: A simple group whose first-order theory is stable in $\aleph_0$

is a simple algebraic group over an algebraically closed field.

The Main Gap conjecture, e.g. for uncountable first order theories, for AECs, and for $\aleph_1$

-saturated models of a countable theory.[94]

Determine the structure of Keisler's order[95][96]

The stable field conjecture: every infinite field with a stable first-order theory is separably closed.

Is the theory of the field of Laurent series over $\mathbb{Z}_p$

decidable? of the field of polynomials over $\mathbb{C}$

?

(BMTO) Is the Borel monadic theory of the real order decidable? (MTWO) Is the monadic theory of well-ordering consistently decida-
[97]

The Stable Forking Conjecture for simple theories[98]

For which number fields does Hilbert's tenth problem hold?

THE EQUATIONS AND SOLUTIONS TO UNSOLVED PROBLEMS

Assume K is the class of models of a countable first order theory omitting countably many types. If K has a model of cardinality $\beth_{\omega_1}$

does it have a model of cardinality continuum?[99]

Shelah's eventual Categority conjecture: For every cardinal λ

there exists a cardinal $\mu(\lambda)$

such that If an AEC K with $LS(K) <= \lambda$

is categorical in a cardinal above $\mu(\lambda)$

then it is categorical in all cardinals above $\mu(\lambda)$

.[94][100]

Shelah's categoricity conjecture for $L_{\omega_1,\omega}$

: If a sentence is categorical above the Hanf number then it is categorical in all cardinals above the Hanf number.[94]

Is there a logic L which satisfies both the Beth property and Δ-interpolation, is compact but does not satisfy the interpolation property? [101]

If the class of atomic models of a complete first order theory is categorical in the $\beth_n$

, is it categorical in every cardinal?[102][103]

Is every infinite, minimal field of characteristic zero algebraically closed? (Here, "minimal" means that every definable subset of the structure is finite or co-finite.)

Kueker's conjecture[104]

Does there exist an o-minimal first order theory with a trans-exponential (rapid growth) function?

Lachlan's decision problem

Does a finitely presented homogeneous structure for a finite relational language have finitely many reducts?

Do the Henson graphs have the finite model property? (e.g. triangle-free graphs)

The universality problem for C-free graphs: For which finite sets C of graphs does the class of C-free countable graphs have a universal member under strong embeddings?[105]

The universality spectrum problem: Is there a first-order theory whose universality spectrum is minimum?[106]

Number theory

THE EQUATIONS AND SOLUTIONS TO UNSOLVED PROBLEMS

General

Grand Riemann hypothesisGeneralized Riemann hypothesisRiemann hypothesis

SOLUTION (to an extent): Could it be that the Riemann Hypothesis is a coherent expression of 'mathematical certainty'? In other words, a 'mathematicized zero'? Maybe woman means 'the inner integral of anti-mathematics'? The thing for which the external is about math? Earlier: Coherent coordinates, then maybe it could be reduced to a logarithm by extending limits. Feminine equations have sex with logarithms? Is that true? (For this result, some credit to Dinko Mehenovik). Trace it on a sphere? See also: Possible Solution to the Riemann Hypothesis

n conjectureabc conjecture

Hilbert's ninth problem

Hilbert's eleventh problem

Hilbert's twelfth problem

Carmichael's totient function conjecture

Erdős–Straus conjecture

Pillai's conjecture

Hall's conjecture

Lindelöf hypothesis

Montgomery's pair correlation conjecture

Hilbert–Pólya conjecture

Grimm's conjecture

Leopoldt's conjecture

Do any odd perfect numbers exist?

Are there infinitely many perfect numbers?

Do quasiperfect numbers exist?

Do any odd weird numbers exist?

Do any Lychrel numbers exist?

Is 10 a solitary number?

Catalan–Dickson conjecture on aliquot sequences

Do any Taxicab(5, 2, n) exist for n > 1?

Brocard's problem: existence of integers, (n,m), such that $n! + 1 = m2$ other than n = 4, 5, 7

Beilinson conjecture

Littlewood conjecture

Szpiro's conjecture

Vojta's conjecture

Goormaghtigh conjecture

Congruent number problem (a corollary to Birch and Swinnerton-Dyer conjecture, per Tunnell's theorem)

Lehmer's totient problem: if $\varphi(n)$ divides $n - 1$,

THE EQUATIONS AND SOLUTIONS TO UNSOLVED PROBLEMS

must n be prime?

Are there infinitely many amicable numbers?

Are there any pairs of amicable numberswhich have opposite parity?

Are there any pairs of relatively primeamicable numbers?

Are there infinitely many betrothed numbers?

Are there any pairs of betrothed numberswhich have same parity?

The Gauss circle problem – how far can the number of integer points in a circle centered at the origin be from the area of the circle?

Piltz divisor problem, especially Dirichlet's divisor problem

Exponent pair conjecture

Is π a normal number (its digits are "random")? [107]

Casas-Alvero conjecture

Sato–Tate conjecture

Find value of De Bruijn–Newman constant

Which integers can be written as the sum of three perfect cubes?[108]

Erdős–Moser problem: is $1^1 + 2^1 = 3^1$ the only solution to the Erdős–Moser equation?

Is there a covering system with odd distinct moduli?[109]

The uniqueness conjecture for Markov numbers[110]

Additive number theory

See also: Problems involving arithmetic progressions

Beal's conjecture

Fermat–Catalan conjecture

Goldbach's conjecture

The values of g(k) and G(k) in Waring's problem

Lander, Parkin, and Selfridge conjecture

Gilbreath's conjecture

Erdős conjecture on arithmetic progressions

Erdős–Turán conjecture on additive bases

Pollock octahedral numbers conjecture

Skolem problem

Determine growth rate of rk(N) (see Szemerédi's theorem)

Minimum overlap problem

Do the Ulam numbers have a positive density?

Algebraic number theory

THE EQUATIONS AND SOLUTIONS TO UNSOLVED PROBLEMS

Are there infinitely many real quadratic number fields with unique factorization(Class number problem)?

Characterize all algebraic number fields that have some power basis.

Stark conjectures (including Brumer–Stark conjecture)

Kummer–Vandiver conjecture

Greenberg's conjectures

Nathan Coppedge

Computational number theory

Integer factorization: Can integer factorization be done in polynomial time?

Prime numbers

Brocard's Conjecture

Catalan's Mersenne conjecture

Agoh–Giuga conjecture

The Gaussian moat problem: is it possible to find an infinite sequence of distinct Gaussian prime numbers such that the difference between consecutive numbers in the sequence is bounded?

New Mersenne conjecture

Erdős–Mollin–Walsh conjecture

Are there infinitely many prime quadruplets?

Are there infinitely many cousin primes?

Are there infinitely many sexy primes?

Are there infinitely many Mersenne primes (Lenstra–Pomerance–Wagstaff conjecture); equivalently, infinitely many even perfect numbers?

Are there infinitely many Wagstaff primes?

Are there infinitely many Sophie Germain primes?

THE EQUATIONS AND SOLUTIONS TO UNSOLVED PROBLEMS

Are there infinitely many Pierpont primes?

Are there infinitely many regular primes, and if so is their relative density $e^{-1/2}$?

For any given integer b which is not a perfect power and not of the form $-4k4$ for integer k, are there infinitely many repunitprimes to base b?

Are there infinitely many Cullen primes?

Are there infinitely many Woodall primes?

Are there infinitely many Carol primes?

Are there infinitely many Kynea primes?

Are there infinitely many palindromic primes to every base?

Are there infinitely many Fibonacci primes?

Are there infinitely many Lucas primes?

Are there infinitely many Pell primes?

Are there infinitely many Newman–Shanks–Williams primes?

Are all Mersenne numbers of prime index square-free?

Are there infinitely many Wieferich primes?

Are there any Wieferich primes in base 47?

Are there any composite c satisfying $2c - 1 \equiv 1 \pmod{c2}$?

For any given integer a > 0, are there infinitely

many primes p such that $a^{p-1} \equiv 1 \pmod{p^2}$? [111]

Can a prime p satisfy $2^{p-1} \equiv 1 \pmod{p^2}$ and $3^{p-1} \equiv 1 \pmod{p^2}$ simultaneously?[112]

Are there infinitely many Wilson primes?

Are there infinitely many Wolstenholme primes?

Are there any Wall–Sun–Sun primes?

For any given integer a > 0, are there infinitely many Lucas–Wieferich primesassociated with the pair (a, −1)? (Specially, when a = 1, this is the Fibonacci-Wieferich primes, and when a = 2, this is the Pell-Wieferich primes)

Is every Fermat number $2^{2^n} + 1$ composite for n>4

?

Are all Fermat numbers square-free?

For any given integer a which is not a square and does not equal to −1, are there infinitely many primes with a as a primitive root?

Artin's conjecture on primitive roots

Is 78,557 the lowest Sierpiński number (so-called Selfridge's conjecture)?

Is 509,203 the lowest Riesel number?

Fortune's conjecture (that no Fortunate number is composite)

Landau's problems

THE EQUATIONS AND SOLUTIONS TO UNSOLVED PROBLEMS

Feit–Thompson conjecture

Does every prime number appear in the Euclid–Mullin sequence?

Does the converse of Wolstenholme's theorem hold for all natural numbers?

PARTIAL SOLUTION: Leudesdorf might have the solution for converse primes, if that's what they are.

Elliott–Halberstam conjecture

SOLUTION: If the lowest proven bound was 248, and the meaning of the counting theorem is essentially decimal relation, a lower bound might be 28 or 136 as these express lower decimal relations.

Problems associated to Linnik's theorem

SOLUTION: My intuition is that the realistic upper bound is exactly 4.

Find the smallest Skewes' number

SOLUTION: About $9/7 * 10\textasciicircum 316$. General formula: lowest hard number (in this case 7) + lowest problem number (in this case 2) divided by lowest hard number (7) * number system ^ (number system - 1) * (next prime * number system + next prime + 1) + number system.

Partial differential equations

Regularity of solutions of Vlasov–Maxwell equations

SOLUTION: epa (f).

Regularity of solutions of Euler equations

SOLUTION: Some clues into Euler here: Euler

Ramsey theory

The values of the Ramsey numbers, particularly R(5,5)

SOLUTION: Probably solved computationally by Yaroslav S here: A 53-Year-Old Network Coloring Conjecture Is Disproved

The values of the Van der Waerden numbers

SOLUTION: Sq root of number of colors multiplied by number of digits. $1 / r^2 = k$, roughly.

THE EQUATIONS AND SOLUTIONS TO UNSOLVED PROBLEMS

Set theory

The problem of finding the ultimate core model, one that contains all large cardinals.

Similar to Woodin's Omega- Hypothesis except ranging over surfaces, logics, and whatever follows next.

If $\beth_\omega$ is a strong limit cardinal, then $2^{\beth_\omega} < \beth_{\omega_1}$ (see Singular cardinals hypothesis). The best bound, $\beth_{\omega_4}$, was obtained by Shelah using his pcf theory.

I tend to agree with the result, a larger version might be lim sum (all = nothing). Might admit of divine exceptions and absolute structural exceptions.

Woodin's Ω-hypothesis.

SOLUTION: This is the problem, if it is a problem, of opposite day, the solution to which essentially is that it is not a problem, or is not opposite, but if it is a problem it has a solutionn unless it is a universal problem, and if it is opposite it cannot be universal unless it is everything. So, the question is can it be the opposite of nothing? But if it is opposite again it is not, and then it is not everything, and then it is nothing, and then it is something else. We don't know that language or mathematics are empirical, the oppositeness might already be incorporated. At that point one might agree there is a kind of mathematical oppositeness, but if so it is only true of non-mathematics, so it must be false. Then it is just the embracing of the arbitrary, so it is defined as arbitrary |

dimension. What is dimensional is not random, what is observed is not arbitrary. Powerful lies hold equivalency, and appear weak. (Note also: the principle that one ought to like what one currently dislikes may be stronger. That is, the law of inclusion or attraction may be superior to the law of difference or contrast. For example, it may be more true that we should be attracted to a different system than to say the system could be contradicted).

(Cont'd)

Rule Against Intermediates

Vs.

The Counting Problem.

Once we add quantum, this might resolve against definite properties that are not observable. —Wheel Problems

Does the consistency of the existence of a strongly compact cardinal imply the consistent existence of a supercompact cardinal?

SOLUTION: No, the triviality of the Continuum Hypothesis [mentioned later] assumes no special distinction unless what is meant is something like the properties of sets. The properties of sets will be at least trivially consistent, and from a naive standpoint cases will always vary.

THE EQUATIONS AND SOLUTIONS TO UNSOLVED PROBLEMS

(Woodin) Does the Generalized Continuum Hypothesis below a strongly compact cardinal imply the Generalized Continuum Hypothesis everywhere?

SOLUTION: Irreversible cardinality would imply yes, in a manner of speaking, but only in some systems which assume irreversibility. That is, if there are proportional numbers, where infinitary numbers may revert to proportional numbers without subtracting or dividing infinity, the continuum hypothesis will eventually look trivial. However, as stated three problems later, proportional numbers may be infinitary probabilities rather than numbers.

Does there exist a Jónsson algebra on $\square\omega$?

SOLUTION: (Answer first). If $\square\omega$ is taken to be complete (in philosophical terms absolute, ultimate, total) and continuous, and infinitary numbers are permitted, then evidently $\square\omega$ admits of no Jónsson cardinals, which indicates no Jónsson algebra. If it's not ultimate, which I think is currently against the definition, then it could if defined that way.

Without assuming the axiom of choice, can a nontrivial elementary embedding V→V exist?

SOLUTION: Yes, by implication of set holes, although we will not know if the set is empty unless we apply the axiom of choice.

Does the Generalized Continuum Hypothesis entail $\Box(E\lambda+cf(\lambda))$

for every singular cardinal λ

?

SOLUTION: No, the proportional numbers are greater than the infinitesimals of infinity, but less than infinity. So, there can be a proportional number for half of infinity. However, proportional numbers are not widely accepted by mathematicians in my understanding, due to the countability problem. The analogy may be similar to probability of infinity. Proportional numbers may place restraints on how number lines can be expressed, for example, they may imply a choice of dimensions or a certain geometric shape. They may be a bridge to variations on trans-finite geometry, particularly geometric divisions of values assumed to be cardinal infinites.

Does the Generalized Continuum Hypothesis imply the existence of an $\Box 2$-Suslin tree?

SOLUTION: Well, if you mean mathematically possible, then yes, as soon as we posit the Continuum Hypothesis or sufficient evidence of the Continuum, they are mathematically possible since the height would be a variation on infinity and infinity is part of the Continuum Hypothesis. Are they countable is a different question, but it is evidently true that infinite number lines are not countable just like the digits of (pi) are not countable. As for whether they are exactly infinity, they are only exactly infinity if we assume one of the dimensions is at least a proportional number (that is, greater than a infinitesimal ratio of one infinity), and if infinity is considered complete.

THE EQUATIONS AND SOLUTIONS TO UNSOLVED PROBLEMS

Is OCA (Open coloring axiom) consistent with
$2\aleph_0 > \aleph_2$

?

SOLUTION: This took some work. If zero is a proof, we assume the search for foundations is abandoned, and coloring requires a standard (maybe a standard qua formalism), and coloring is always formal if we assume mathematics is formal and that zero is mathematical and the only foundation. If there are other foundations there may be multiple distinct proof theories. If zero is not proof, nature or infinity provide the solution no matter how simple, and/or proof lies outside that concept of mathematics or within a concept at least as foundational as zero, or foundations are not preferred for proof, and / or the search for foundations is not over.

Assume ZF and that whenever there is a surjection from A onto B there is an injection from B into A. Does the Axiom of Choice hold?[113]

SOLUTION: Although it is tempting to say that the only exception is illusory sets, in fact at least one other counterexample exists with exponential classification. It is easy to imagine that multiple classifications would eventually prove examples that don't exist at all, and the mere existence of classification may imply a kind of error of obsession. How can we find objects without then finding infinite objects? And if we know objects aren't infinite, how do we know they exist? But if we don't find infinite objects, how can we say we know how to count? After all, if we don't count infinite objects, we might not have evidence for the reality of numbers as a system. How do we disprove exponential classification in more than one dimension, if the area of a plane requires exponents? Doesn't this imply knowing the numbers of dimensions? Key point: If we don't know the number of dimensions there may be no way to classify without overlapping sets.

Topology

Borel conjecture

SOLUTION: In 4-d, if 3-d axes divide symmetrically, the cells may be symmetric in 4-d, but not in 3-d. This is because anything 3-d becomes possible if it is ambiguous in 4-d. This is analogous to hyperbolic-isometric space, such as moving past trees with a speed of zero. It is very challenging up to dimension 3.5, however illusions of it exist in 2-d and 3-d with analogy to shadows, holograms, and noise.

Hilbert–Smith conjecture

SOLUTION: Unless the graph is symmetric, any formal symmetry will produce an asymmetric response to the formalism. Equivalently, if the Reimann structure is not formally coherent it will be formally asymmetric or coherence has been disregarded even if the map, etc is symmetric in any way. These views could also conceivably be interpreted mathematically in terms of completeness, as if to say a Reimann structure must be completely incomplete to qualify as mathematics, or the formal asymmetry is expressible mathematically.

THE EQUATIONS AND SOLUTIONS TO UNSOLVED PROBLEMS

Novikov conjecture

SOLUTION: Of course, such invariants are partial to scale, less so when constrained by d. A solution may not be expressible mathematically because of conceptual ambiguity. It is something like a sine operation on cohomology classes, which goes well with the solution to homological conjectures in commutative algebra [mentioned as the first problem, which was very hard].

Unknotting problem

SOLUTION: Express cubically. Grasp two opposite diagonal corners. If all lobes on both sides can be folded through other lobes, it is solvable. If this can be done in polynomial time, then it is solvable in polynomial time. Essentially, if lobes are even in number and evenly spaced when unfolded, e.g. because all even spaces will produce a single twist unless the number of lobe overlaps is odd.

Whitehead conjecture

SOLUTION: In isometric or warped such as symmetric relativistic space Whitehead can be false because inner morphisms, although usually asymmetric in this problem, do not always match outer.

Zeeman conjecture

SOLUTION: A Y-axis measure is of course collapsible by relation, implying the POSSIBILITY of time-equivalence, although not always universally. The remaining question is then one of possibility, governed formally or otherwise.

Volume conjecture

SOLUTION: Something related to $3.5d^n$, where n is symmetry or colors.

...

THE EQUATIONS AND SOLUTIONS TO UNSOLVED PROBLEMS

MISCELLANEOUS SOLUTIONS:

Attempt to Solve the Collatz Conjecture

50% chance of dividing by 2.

50% chance of multiplying by 4 - 25% + 1.

Add it together:

50% - 25% = 25%

4/2 = 2

X * 2 * 0.25 + 1... must approach 1 or the conjecture is false.

So, 0.5X + 1 must approach 1 or the conjecture is false.

That's simple, because X can be a negative number, X averages to zero, snd the result tends to be the added number.

Why doesn't someone try it for:

Evens / 2

Odds X 3 + 2.

The result will probably be 2 or a difference in terms of 0.5X.

Sofa Problem [was from Sam Wise 144 on Quora however it appears to have been deleted unfortunately. The address was: https://www.quora.com/What-is-the-most-interesting-math-problem-you-have-seen-recently/answer/Sam-Wise-144/comment/132019205)

How can I calculate the probability of something happening twice or thrice?

According to the Theory of Everything, (Set 0 > Efficiency + Difference)...

If the thing is acted on, and the number of times is Set 0 = X, the coherent probability in this case is the inverse of the impossibility of X. So, if the impossibility of X is < X there is some probability, and if impossibility = 0, then probability = 1.

This is not traditional probability however, but there may be some compatibility at a stretch. In the case of impossibility of X < X it is simply saying that all counted elements of X must be present to have full probability. In the case of dice, a Theory of Everything value of 2 would atill translate as a probability of 1/2, or 2 out of something depending on the context. In some cases values would be subsets of the efficiency, with the impossibility held separate.

Thus, in any case:

Acted on = 1 / (Value 1 + value 2)... + Impossibility.

Acting = |Set 0 + - impossibility| + impossibility

THE EQUATIONS AND SOLUTIONS TO UNSOLVED PROBLEMS

Possible Solution to the Riemann Hypothesis

Big Update:

In my recent attempts:

$(1/0.75)(d-1)(d/d^2)$

Where d is dimensions, produces progression:

(4-d =) 1, (8-d =) 1.16666, (16-d=) 1.25.

These are multipliers of some type related to sq root of 0.5

This emerged from my earlier work on tthe theory of everything and is ongoing.

Another formula:

$(d*3.333) - |(+2/d) + 0.5|$

PREVIOUS WORK, NOT NECESSARILY IN A COMPLETE STATE:

If 1/2 is True, all values lie at 1/2 + it where t is a real number.

There are only 2 ways to get 0.5...

(WORKING)

$2/D >$ results / verbs $= 1/|+t| = 0.5$

$2/D >$ results / verbs $= 1/|-t| = 0.5$

As in this problem

$1/|t| = 0.5$

This gives a situation where the most objective number is +/- 2 modified by 1 / the absolute value.

4 / 2^3

8 / 2 root of modulo 4.

0.25t^2 * 8

Note: This solution may require some modification.

Alternatives to the given expression which solve 0.5 may be solutions to the Riemann Hypothesis:

Link: Objective: Knowledge

0.70711

(Results / Verbs) / .70711 = Modifier * 0.3535

NEWER:

(1/0.75)(d -1)(d / d ^2)

Where d is dimensions, produces progression:

1, 1.16666, 1.25

THE EQUATIONS AND SOLUTIONS TO UNSOLVED PROBLEMS

EARLIER CHART OF KNOWLEDGE ONLY MILDLY RELATED

2D 1 result / 2 verbs * 0.70711 = 0.3535...

2D 2 results / 4 verbs * 0.70711 = 0.3535...

2D 4 results / 8 verbs * 0.70711 = 0.3535...

2D 6 results / 12 verbs * 0.70711 = 0.3535...

2D 8 results 16 verbs ... etc

4 D 1 result / 2 verbs ...

8 D 6 results / 8 verbs... * 0.707110 = 0.5303325

... 6 / 8 * .3535

8 D 12 results 16 verbs... etc.

8 D 15 results 20 verbs... etc.

16 D 5 results 8 verbs etc.

16 D 10 results 16 verbs

16 D 15 results / 24 verbs * 0.70711 = 0.44194375

Which do you prefer, Von Neumann's version of set theory or Zermelo's? Why?

From what I understand, Zermelo set theory is an attempt to organize which does not organize, and Von Neumann's set theory is an attempt to make sets that does not make sets.

We can see immediately that the two are not methodologically compatible, or at least not independently. I can see they are compatible in computing, which is an interesting fact.

But it makes me wonder, are they both—in light of Von Neumann's— merely designed as a means to an end? And if so, how can they be viewed objectively? Is that a feint at Quantum Theory, and what would separate definitions even mean if the whole works is vertically substitutional *physically?*

Von Neumann's logic does claim to be substitutional (I'm guessing), in that certain sense of Von Neumann's, but that doesn't make it objectively quantum if there were such a thing would it?

(Recent articles on Quantum Darwinism remind me of Von Neumann's theory, but it does not seem to break up well. Neumann's set theory is like an attempt at coherent Quantum Darwinism: a noble task, but doomed from every average perspective).

However, the implication of both in computing suggests that quantum is what was meant to be expressed by either theory in light of Von Neumann—which could even

THE EQUATIONS AND SOLUTIONS TO UNSOLVED PROBLEMS

suggest that Neumann's theory is physically wrong.

Now, why would we have to elaborate end-over-end with plain normative logic just to reach the conclusion that Von Neumann is wrong?

That is very frustrating, and if there is any truth in this both of these set theories have fundamental flaws.

But if that is the case, we are just saying we want a completely separate methodology—and they are good for what they are.

But maybe arbitrary is good after all. However, that doesn't sound like either theory.

It is like combined together they are arbitrary in just such a way as to be inconveniently non-arbitrary.

Why not just say, 'random event X happened' let's research it and build something on this, instead of, we have observed Y events and we think it points to Set Theory A or B?

There is something coherently flawed with adopting a model which does not express it's criteria.

Still, for computing it seems Von Neumann's theory has huge advantages which have no existence in Zermelo's theory. Actually applying those advantages has proven too unconventional for the most part I would guess.

I have trouble grasping why Von Neumann's theory even looks like mathematics. Doesn't it place very many restraints on how the behavior is expressed?

I would almost suggest starting with something more preschool and simply laying out fundamentals as separate contingent facts, and just structure any further theories on the degree of fundamentality. However, this clearly is not science this way, or not set theory science.

It seems once again that if science is to be coherent it requires a formula.

As far as requiring a formula, talk of fundamental philosophy may look like water under the bridge, not because philosophy is unimportant, but because philosophy is prone to think fundamentals are sufficient evidence for anything.

Hilariously it looks like in his own way even Neumann was guilty of that. Which, some may say, applies to every theory of sets ever conceived.

The fundamentals are philosophical is unavoidable, but if we look for some type of natural law, axioms will almost certainly be secondary because axioms are set theory not evidence.

For reasons like this 'coherent' versions of set theory would look for axioms that apply universally to observed phenomena by using logic which is somehow capable of describing everything. In this case, describing all of mathematics looks like the wrong move, because mathematics isn't literally everything.

(Presupposing that something is mathematics because it looks like mathematics runs the course of reducing to mathematics, which may as well be assuming the consequence, which we know can be filled in with all sorts of arbitrary content).

Now the question is, is a set theory coherent? Well, mathematics argues it can't be by referring implicitly to empirical phenomena, but by referring to empirical phenomena as mathematics, set theory is already making a fundamental mistake. I would prefer to structure based on the degree of fundamentality: a pleuroverse of ideas, in which theories account for diverse aspects of logic, mathematics, and experience.

We can easily see from this vantage point that set theory has very little to do with physics, except that the fundamentals of physics and the fundamentals of axioms are held together by the common notion of philosophy. At THIS point we can form a set theory which is fundamental, but if so we must accept that it is arbitrary yet fundamental, for if it were not our hands would be tied to the idea that fundamentals produce math from which are produced fundamentals, which is clearly circular.

In an arbitrary world we can say there is an end to counting because everything is assessment, and assessment is sufficient. And if we assess that something is arbitrarily fundamental, or assessment is that it is fundamental and that is sufficient. But if not, what is the real scale? What makes an axiom important? How can we claim our axioms are complete or that some special rule does not apply? If science is fundamental it must be philosophy, or it must have a formula, while the formula scientifically will refer to nothing philosophically fundamental because that would imply lack of critique, and if it were philosophy it could be expressed only as axioms which must themselves be fundamental and so inherently un-scientific, since scientifically the goal of axioms is to build mathematics, leading to circular reasoning.

So, my understanding is the solution is arbitrary fundamentality, but this implies no inherent negative feelings. It does not refute science for what it's worth, it simply says there is something fundamental which we can choose to notice, and when we choose to notice it the result happens to be arbitrary, because arbitrary is a real standard of logic.

However, if we do not pursue coherence, the results are not coherent in spite of any level of understanding we might apply.

Key to the X Axis

Square Peg Problem: Square peg problem (December 17, 2020): Three best related solutions: simple solution is if there are two parallel lines, many squares form in any location where the length along the lines is equal to the distance between the lines. To be more complex, for any two points, are there two points perpendicular at the same distance as the length between the original points? If not, pick original two points that are closer together on the same side if the opposite side is at shorter distance than before, and further apart if the opposite side is at longer distance than before.

Square Peg Problem: Square peg problem (December 17, 2020): Three best related solutions: simple solution is if there are two parallel lines, many squares form in any location where the length along the lines is equal to the distance between the lines. To be more complex, for any two points, are there two points perpendicular at the same distance as the length between the original points? If not, pick original two points that are closer together on the same side if the opposite side is at shorter distance than before, and further apart if the opposite side is at longer distance than before.

Nathan Coppedge

SOLUTIONS TO UNSOLVED PROBLEMS IN NEUROSCIENCE

SOLVED ALL THESE PROBLEMS IN ABOUT AN HOUR, WHEN I GOT TO IT…

PROBLEMS TAKEN FROM WIKIPEDIA: List of unsolved problems in neuroscience - Wikipedia

(Real date: OCT 28, 2018)

PROBLEM:

Consciousness: What is the neural basis of subjective experience, cognition, wakefulness, alertness, arousal, and attention? Is there a "hard problem of consciousness"? If so, how is it solved? What, if any, is the function of consciousness?

Subjective experience = Objective limits on perception. Question of manifesting intelligence is the answer of not manifesting limited intelligence, so it is an adaptation designed to create a result of intelligence, probably arising from an unintelligent context. Alertness is cognition arising from un-alert things, for example, prey, or it would have no meaning, so it could be for hunting. Arousal is obviously a response to the condition of sleep. If we say consciousness has a hard problem, it doesn't have a soft solution, and if it is likely it is not likely to involve consciousness. So, solutions to the hard problem of consciousness are unlikely solutions involving consciousness, and likely solutions not involving consciousness.

Or, simply, luck (best unlikely theory) involving consciousness, or commonality or shared traits (best likely theory) involving unconsciousness.

PROBLEM:

Perception: How does the <u>brain</u> transfer <u>sensory</u> information into coherent, private percepts? What are the <u>rules</u> by which perception is organized? What are the features/<u>objects</u> that constitute our perceptual experience of internal and external events? How are the <u>senses</u> integrated? What is the relationship between subjective experience and the <u>physical</u> world?

The brain transfers sensory information into coherent, private perceptions through bodily primitive reactions out of incoherent, public perceptions. The rules by which perception is organized is by fiat, arbitrarily, in short, not organized. Internal events try to order external events, just as external events try not to influence internal events. The senses are chemical reactions disintegrating, the rational purpose was imposed through organization, which is a kind of extraneous characteristic, an offshoot of chemical reactions disintegrating. The relationship between subjective experience and the external world is that it is objective and internal.

PROBLEM:

Learning and **memory**: Where do our memories get stored and how are they retrieved again? How can learning be improved? What is the difference between **explicit** and **implicit** memories? What molecule is responsible for **synaptic tagging**?

Our memories get stored when we forget to forget them, and it involves things which are not memories: blunt sensations. Learning can be improved by shutting up and listening. The difference between explicit and implicit memories, is that explicit memories are implied to be brutal facts, whereas implicit memories are obvious brutal facts. The molecule responsible for synaptic tagging is the molecule irresponsible and against disintegralized quanta, in other words, the careless, stupid, disoriented, and distracted one, or the authentic one.

…

THE EQUATIONS AND SOLUTIONS TO UNSOLVED PROBLEMS

PROBLEM:

Neuroplasticity: How **plastic** is the mature brain?

The quantity of mature plasticity question is answered by a quality of immature hardening.

PROBLEM:

Development and **evolution**: How and why did the brain **evolve**? What are the **molecular** determinants of individual brain development?

The question of the brain is answered by 'why not be immature?' The quality of individual brain substances are determined by an indeterminate number of immature groups.

PROBLEM:

Free will, particularly the **neuroscience of free will**

The freedom of the sensory system is determined by reactions against systems.

PROBLEM:

Sleep: What is the biological function of sleep? Why do we **dream**? What are the underlying brain mechanisms? What is its relation to **anesthesia**?

The biology of sleep is that we die if we don't sleep, and we can be functional or dysfunctional about this. The experience of dreams is from inexperience with not being a visionary, in other words, curiosity about thinking and visual vision. The curious brain (dreaming brain) arises from lack of curiosity about the body. Physical knock-out drugs affect sleep through non-abstract effects that dose-up the chemical system.

THE EQUATIONS AND SOLUTIONS TO UNSOLVED PROBLEMS

PROBLEM:

Cognition and **decisions**: How and where does the brain evaluate **reward** value and effort (**cost**) to modulate **behavior**? How does previous experience alter perception and behavior? What are the genetic and environmental contributions to brain function?

The brain location of costs-to-benefits is embodied with qualities of double-negative benefits, this has the result of creating brain plasticity. Past experiences result in behavior that is future-oriented, disoriented, and well-behaved.

PROBLEM:

Language: How is it implemented neurally? What is the basis of semantic meaning?

The neurology of languages is chemical, and dis-associative. People learn things that have no negative association, or they forget things they feel happy about. For example, a son of divorced parents may be a schizophrenic about language. A person with early depression may feel words make them happy. Someone who is bullied using words will find words to be pleasantly tactile, while someone abused physically will find words are pleasantly abstract. However, if they feel very happy, or if they are forced to associate, they may fail to learn proper language. Semantic meaning is the ability to label meanings, in other words, it is the attachment to the word 'semantic' as opposed to the 'un-semantic'.

PROBLEM:

Diseases: What are the neural bases (causes) of **mental** diseases like psychotic disorders (e.g. **mania**, **schizophrenia**), **Amyotrophic lateral sclerosis**, **Parkinson's disease**, **Alzheimer's disease**, or **addiction**? Is it possible to recover loss of sensory or motor function?

- **One possible neural basis for mental illness is law integration disorder, a term coined by Johan Nygren, a Swedish gentleman who studied medicine and gene memes. While things like Parkinsons are easily explained medically by lack of dopamine, probably caused by environmental epigenetic effects like exposure to pesticides, other mental disorders lack a physiological basis. The theory of law integration disorder factors in the fight or flight response to coercion and explains these non-physiological mental illnesses as stress-induced phenomena.**

The neurological origin of mental disease is the bodily or disintegral destination of a healthy body, in other words, concern for one's physical health under the constraint of linear time. The condition of being quantum-God, quantum-brain, or quantum-schizophrenic. Bad options resulting from good determinations.

PROBLEM:

Movement: How can we move so controllably, even though the motor nerve impulses seem haphazard and unpredictable?

Control of movement explanation involving physical atoms is compulsion, possibly involving virtual particles.

PROBLEM:

Computational theory of mind: What are the limits of understanding thinking as a form of computing?

Logic gates as mind theory is relevant in the sense that there is an irrationality that is blocked, part of the body, not theoretical, but it is not necessarily relevant.

PROBLEM:

Computational neuroscience: How important is the precise timing of action potentials for information processing in the neocortex? Is there a canonical computation performed by cortical columns? How is information in the brain processed by the collective dynamics of large neuronal circuits? What level of simplification is suitable for a description of information processing in the brain? What is the neural code?

The neural code is the bodily or disintegralized independence, identity, improbability. The cause of the neural code is that it embodies a solution to codes, or more precisely it requires a code.

PROBLEM:

How does **general anesthetic** work?

Knock-out drugs of no description work by dosing up the chemical system of some kind.

PROBLEM:

Neural computation: What are all the different types of neuron and what do they do in the human brain?

The question of the many kinds of neurons is answered by the singular non-typed genera of the body.

PROBLEM:

Noogenesis - the emergence and evolution of intelligence: What are the laws and mechanisms - of new idea emergence (insight, creativity synthesis, intuition, decision-making, eureka); development (evolution) of an individual mind in the ontogenesis, etc.?

New synapses are formed from the obvious deformation of the disintegral.

...

SOLUTIONS TO UNSOLVED PROBLEMS IN PHILOSOPHY

METAPHYSICS:

PROBLEM: Why there is something rather than nothing

Main article: Problem of why there is anything at all

The question about why is there anything at all instead of nothing has been raised or commented on by philosophers includingGottfried Wilhelm Leibniz, [12]

The question is general, rather than concerning the existence of anything specific such as the universe/s, theBig Bang, mathematical laws, physical laws,
time, consciousness or God.

SOLUTION: Everything specific exists how? Problems are more complex and difficult, you need to create something to have a problem, and general problems require more and more evidence to have any problem, so solutions always out-weigh problems unless all problems are general or specific and all solutions are the opposite specific or general. In short, all true general solutions solve all specific problems, and all true specific solutions solve all general problems, therefore problems are exponentially solved or there is a problem with truth.

PROBLEM: Universals, substances described as having common properties.

SOLUTION:

Idea representation quandary. We should not consider thoughtlessness that is not represented. Thoughts have reality too. There is a common reality in which thoughts aren't everything, after all we cannot change Dr. Johnson's rock, and if everything is thought then there is no problem to the idea that things have qualities.

PROBLEM:

Principle of individuation

Main article: Principle of individuation

Related to the quarrel of universals, the principle of individuation is what individuates universals.

SOLUTION:

Natural categories unexplained. The explanation is categories are natural, there is an explanation, provided by nature. An explanation was desired or categories may be hard to explain. General problems have specific solutions, because opposites have alternatives in their opposites and everything else is neutral.

THE EQUATIONS AND SOLUTIONS TO UNSOLVED PROBLEMS

PROBLEM: Sorites paradox

Otherwise known as the "paradox of the heap", the question regards how one defines a "thing." Is a bale of hay still a bale of hay if you remove one straw? If so, is it still a bale of hay if you remove another straw? If you continue this way, you will eventually deplete the entire bale of hay, and the question is: at what point is it no longer a bale of hay? While this may initially seem like a superficial problem, it penetrates to fundamental issues regarding how we define objects. This is similar to Theseus' paradox and the Continuum fallacy.

SOLUTION:

Quantity of categorical completeness quandary. there is no question that the quality of being un-categorical is incomplete. Measurement of a category implies a quantity, unless we are being vague or have the wrong idea. It is very possible we have the wrong idea of category.

Nathan Coppedge

PROBLEM:

Also known as the **Ship of Theseus,** this is a classical paradox on the first branch of metaphysics, Ontology (philosophy of existence & identity). The paradox runs thus: There used to be the great ship of Theseus which was made out of, say, 100 parts. Each part has a single corresponding replacement part in the ship's storeroom. The ship then sets out on a voyage. The ship sails through monster-infested waters, and every day, a single piece is damaged and has to be replaced. On the hundredth day, the ship sails back to port, the voyage completed. Through the course of this journey, everything on the ship has been replaced. So, is the ship sailing back home the ship of Theseus or no?

If yes, consider this: the broken original parts are repaired and re-assembled. Is this the ship of Theseus or no? If no, let us name the ship that sails into port "The Argo". At what point (during the journey) did the crew of the Theseus become the crew of the Argo? And what ship is sailing on the fiftieth day? If both the ships trade a single piece, are they still the same ships?

This paradox is a minor variation of the Sorites Paradox above, and has many variations itself. Both sides of the paradox have convincing arguments and counter-arguments, though no one is close to proving it completely.

THE EQUATIONS AND SOLUTIONS TO UNSOLVED PROBLEMS

SOLUTION:

Serious alteration of identity equal to the same question. Un-seriously it's the same difference, seroously it's not identical but may serve the same function.

PROBLEM: **Material implication**

Main article: <u>Material conditional</u> People have a pretty clear idea what if-then means. In <u>formal logic</u> however, material implication defines if-then, which is not consistent with the common understanding of conditionals. In formal logic, the statement "If today is Saturday, then 1+1=2" is true. However, '1+1=2' is true regardless of the content of the antecedent; a causal or meaningful relation is not required. The statement as a whole must be true, because 1+1=2 cannot be false. (If it could, then on a given Saturday, so could the statement). Formal logic has shown itself extremely useful in formalizing argumentation, philosophical reasoning, and mathematics. The discrepancy between material implication and the general conception of conditionals however is a topic of intense investigation: whether it is an inadequacy in formal logic, an ambiguity of ordinary <u>language</u>, or as championed by <u>H.P. Grice</u>, that no discrepancy exists.

SOLUTION:

Gettier Problems true if implied does not imply true. You need to have a standard to have truth.

Nathan Coppedge

SOLUTIONS TO LONG-STANDING PROBLEMS IN PHILOSOPHY

PHILOSOPHY OF MIND

PROBLEM:

Mind–body problem

The mind–body problem is the problem of determining the relationship between the human body and the human mind. Philosophical positions on this question are generally predicated on either a reduction of one to the other, or a belief in the discrete coexistence of both. This problem is usually exemplified by Descartes, who championed a dualistic picture. The problem therein is to establish how the mind and body communicate in a dualistic framework. Neurobiology and emergence have further complicated the problem by allowing the material functions of the mind to be a representation of some further aspect emerging from the mechanistic properties of the brain. The brain essentially stops generating conscious thought during deep sleep; the ability to restore such a pattern remains a mystery to science and is a subject of current research (see also neurophilosophy).

SOLUTION: Defined mind-body dependence question, the answer is undefined, independent mutual causality. A mind-body paradigm in which they are commutually related. The problem has not been defined in a solvable way, because bodies could be abstract and minds could be physical. The answer is, if they are related, they are related the way they are related. If we want a relation we must define it. If the person does not define the relation it must be something else which defines it. We can't prove omniscience just by solving a particular problem, particularly one this vague. For some the mind will be in control of a physical identity, for others the body will be in control of an abstract identity, or some other combination. It is partly a matter of definition and can depend on the words we use. Perhaps we need a better concept than 'body' and 'mind' but this might change how we operate. It is contingent on definition and definitions of paradigmatic functioning. What works undemiably works unless it is limited, but this does not guarantee that we have phrased the problem correctly. And, the problem gives no ground for phrasing itself correctly. Basically, it's relative, because intelligence is more important than the body and reality is more important than the mind. So, 'real intelligence' is all that is meant by mind-body duality, unless there us something more exyreme to compare it to. Real intelligence could undoubtedly mean many different things.

PROBLEM:
Cognition and AI

This problem actually defines a field, however its pursuits are specific and easily stated. Firstly, what are the criteria for <u>intelligence</u>? What are the necessary components for defining <u>consciousness</u>? Secondly, how can an outside observer test for these criteria? The "<u>Turing Test</u>" is often cited as a prototypical test of intelligence, although it is almost uni-

versally regarded as insufficient. It involves a conversation between a sentient being and a machine, and if the being can't tell he is talking to a machine, it is considered intelligent. A well trained machine, however, could theoretically "parrot" its way through the test.
This raises the corollary question of whether it is possible to <u>artificially create consciousness</u>(usually in the context of<u>computers</u> or <u>machines</u>), and of how to tell a well-trained mimic from a sentient entity.

Important thought in this area includes most notably: <u>John Searle</u>'s <u>Chinese Room</u>, <u>Hubert Dreyfus</u>' non-cognitivist critique, as well as<u>Hilary Putnam</u>'s work on <u>Functionalism</u>.

A related field is the <u>ethics of artificial intelligence</u>, which addresses such problems as the existence of moral personhood of AIs, the possibility of moral obligations *to* AIs (for instance, the right of a possibly sentient computer system to not be turned off), and the question of making AIs that behave ethically towards humans and others.

SOLUTION:

A.I. A type of intelligence must be of it's type, and it must be of it's type before it is successful, or it cannot be it's type. Success falls in categories. Intelligence must be sensory before it is sensory intelligence, before it is successful sensory intelligence. If sensory intelligence is required for intelligence, a robot will need real senses before it acquires intelligence. This does not mean AI's have the same senses as humans, or that senses are even necessarily required, thusthe hard problem. The core problem is robots cannot be humans, they are categorically different unless

THE EQUATIONS AND SOLUTIONS TO UNSOLVED PROBLEMS

they are merely used as human enhancers. In this way we can define the successful cases as being 'extremely different from humans'.

PROBLEM:

Hard problem of consciousness

The hard problem of consciousness is the question of what consciousness is and why we have consciousness as opposed to beingphilosophical zombies. The adjective "hard" is to contrast with the "easy" consciousness problems, which seek to explain the mechanisms of consciousness ("why" versus "how", or final causeversus efficient cause). The hard problem of consciousness is questioning whether all beings undergo an experience of consciousness rather than questioning the neurological makeup of beings.

SOLUTION:

Hard problem of consciousness: We are atttacted to explanations because we are thinking things. The explanation of consciousness is that there is no explanation for being unconscious. We happen not to know alternatives to being aware —it happens to be within our power if we are conscious as soon as we assume we have power—unless there is something which prevents us from realizing awateness. But, since awareness is fundamental, only the worst things would motivate us to seek unawareness. Therefore, consciousness is desirable and ever since we have gained realization we have sought more of it, like a queen bee eating royal jelly. There is a parity between the desire for awareness and the desire for an explanation. In effect, if for practical or less practical means we became more interested in awateness, we would become more aware, unless, for practical or other reasons we did not desire awareness. Practicality is the hard limit of consciousness: most people have too much doubt and negative messages to achieve super consciousness as humans define it. But 'good messages' and 'curiosity about life' really have parity with consciousness, because they lead to perceptive confrontations. And, at the same time, many people have similarly intense fundamental awareness (say, even if there were levels of consciousness, at every level) because the consciousness insight is itself a large part of the realization of consciousness. Even so, in my experience, there may sonetimes be external limits on consciousness represented by practicality as opposed to wish fulfillment. There is no rule which says that consciousness is the best thing under every definition, it is just somewhat desirable, but not automatically. Many other things may be automatic-ins or more highly desirable, but some of them may involve consciousness or something similar. In Leibniz's theory everything is really conscious, it is just less or more God-like.

THE EQUATIONS AND SOLUTIONS TO UNSOLVED PROBLEMS

SOLUTIONS TO LONG-STANDING PROBLEMS IN PHILOSOPHY

ETHICS

Problem: Moral Luck

Moral luck

The problem of moral luck is that some people are born into, live within, and experience circumstances that seem to change their moral culpability when all other factors remain the same.

For instance, a case of *circumstantial moral luck:* a poor person is born into a poor family, and has no other way to feed himself so he steals his food. Another person, born into a very wealthy family, does very little but has ample food and does not need to steal to get it. Should the poor person be more morally blameworthy than the rich person? After all, it is not his fault that he was born into such circumstances, but a matter of "luck".

A related case is *resultant moral luck.* For instance, two persons behave in a morally culpable way, such as driving carelessly, but end up producing unequal amounts of harm: one strikes a pedestrian and kills him, while the other does not. That one driver caused a death and the other did not is no part of the drivers' intentional actions; yet most observers would likely ascribe greater blame to the driver who killed (compare consequentialism and choice).

The fundamental question of moral luck is how our moral responsibility is changed by factors over which we have no control.

Solution:

Some people are ethically lucky. Not everyone is unethically unlucky: some people could use ethical luck, this is why some people look ethically unlucky.

Problem:

Moral knowledge

Are moral facts possible, what do they consist in, and how do we come to know them? Rightness and wrongness seem strange kinds of entities, and different from the usual properties of things in the world, such as wetness, being red, or solidity. Richmond Campbell

[8]

has outlined these kinds of issues in his encyclopedia article <u>Moral Epistemology</u>.

In particular, he considers three alternative explanations of moral facts as: theological, (supernatural, the commands of God); non-natural (based on intuitions); or simply natural properties (such as leading to pleasure or to happiness). There are cogent arguments against each of these alternative accounts, he claims, and there has not been any fourth alternative proposed. So the existence of moral knowledge and moral facts remains dubious and in need of further investigation. But moral knowledge supposedly already plays an important part in our everyday thinking, in our legal systems and criminal investigations.

SOLUTION:

THE EQUATIONS AND SOLUTIONS TO UNSOLVED PROBLEMS

Ethical Knowledge: Repugnance to not explain the meaningless desire to explain meaning. (If sin is not the antithesis of meaningful ethics, then it isn't really sin. And ethics is not really ethics unless it is meaningful ethics. Therefore, meaning is central to ethics, and meaning concerns in this case an ability to explain. What is sinful essentially is the unexplained unless the unexplained is meaningful, such as meaningful ethics, for example enforced ignorance against fashionable depression. And also, the circumstances of explaining may come into play. What is a meaningful excuse? And, what destroys meaning not just for ethics, but for everyone?

For the philosophical solutions particularly in ethics, some but not all credit is due to Richard Volkman, professor at Southern CT State University. Some thought was involved in remembering his solutions to some of the problems. (Also, a form of analytic method was applied).

Nathan Coppedge

SOLUTIONS TO LONG-STANDING PROBLEMS IN PHILOSOPHY

EPISTEMOLOGY

Gettier problem

Main article: Gettier problem

Plato suggests, in his *Theaetetus* (210a) and*Meno* (97a–98b), that "knowledge" may be defined as justified true belief. For over two millennia, this definition of knowledge has been reinforced and accepted by subsequent philosophers. An item of information's justifiability, truth, and belief have been seen as the necessary and sufficient conditions for knowledge.

In 1963, Edmund Gettier published an article in the journal "Analysis", a peer reviewed academic journal of philosophy, entitled "Is Justified True Belief Knowledge?" which offered instances of justified true belief that do not conform to the generally understood meaning of "knowledge." Gettier's examples hinged on instances of epistemic luck: cases where a person appears to have sound evidence for a proposition, and that proposition is in fact true, but the apparent evidence is not causally related to the proposition's truth.

In response to Gettier's article, numerous philosophers... have offered modified criteria for "knowledge." There is no general consensus to adopt any of the modified definitions yet proposed. Finally, if infallibilismis true, that would seem to definitively solve the Gettier problem for good--the idea is that knowledge requires certainty, such that, certainty is what serves to bridge the gap so that we arrive at knowledge, which means we would have an adequate definition of knowledge. However, infallibilism is rejected by the overwhelming majority of philosophers/epistemologists, even though it would solve the Gettier problem (if true).

SOLUTION:

Gettier Problem: Radically Perpendicular or Contingent Valid Corollary. Semantical Coherence is Not Valid Proof, e.g. the question of if something is valid does not make it valid: validity is a conjecture which must be supported under every relevant standard: only absolute proofs will be absolutely true: Aristotle is not absolute. Coppedge's categorical deduction is absolute in one degree within assumptions. Please analyze well if you decide to use that technique, it is not superficial, but it treats all truths the same.

PROBLEM: Problem of the criterion

Overlooking for a moment the complications posed by Gettier problems, philosophy has essentially continued to operate on the principle that knowledge is justified true belief. The obvious question that this definition entails is how one can know whether one's justification is sound. One must therefore provide a justification for the justification. That justification itself requires justification, and the questioning continues interminably.

The conclusion is that no one can truly have knowledge of anything, since it is, due to this infinite regression, impossible to satisfy the justification element. In practice, this has caused little concern to philosophers, since the demarcation between a reasonably exhaustive investigation and superfluous investigation is usually clear.

Others argue for forms of coherentistsystems, e.g. Susan Haack. Recent work byPeter D. Klein [1] views knowledge as essentially defeasible. Therefore, an infinite regress is unproblematic, since any known fact may be overthrown on sufficiently in-depth investigation.

SOLUTION:

Criterion Problem: Possibility of incoherent justification. It may not be coherently justified.

PROBLEM:

The Molyneux problem dates back to the following question posed by William Molyneux to John Locke in the 17th century: if a man born blind, and able to distinguish by touch between a cube and a globe, were made to see, could he now tell by sight which was the cube and which the globe, before he touched them? The problem raises fundamental issues in epistemology and the philosophy of mind, and was widely discussed after Locke included it in the second edition of his *Essay Concerning Human Understanding*.

SOLUTION:

Mollyneux Problem: Colorblind scientist: inexperienced knowledge with the unknown — experienced with not knowing what one knows. Reset the goal posts, if one can't do it the right way, one is out of luck. The more experience one has with metaphors and sensory experiences, the more one can overthink it, but anyone who perceives something perceived it in some way at some point.

PROBLEM:

The **Münchhausen trilemma**, also called **Agrippa**'s trilemma, purports that it is impossible to prove any certain **truth** even in fields such as logic and mathematics. According to this argument, the proof of any theory rests either on **circular reasoning**, infinite regress, or unproven **axioms**.

SOLUTION:

M Trilemma: Ill-composed faith, well-composed assumptions. Or also, conjecture leads to composition.

PROBLEM:

Qualia. See also: **Distinguishing blue from green in language**

The question hinges on whether **color** is a product of the mind or an inherent property of objects. While most philosophers will agree that color assignment corresponds to spectra of light **frequencies**, it is not at all clear whether the particular psychological phenomena of color are imposed on these visual signals by the mind, or whether such **qualia** are somehow naturally associated with their **noumena**. Another way to look at this question is to assume two people ("Fred" and "George" for the sake of convenience) see colors differently. That is, when Fred sees the sky, his mind interprets this light signal as blue. He calls the sky "blue." However, when George sees the sky, his mind assigns green to that light frequency. If Fred were able to step

THE EQUATIONS AND SOLUTIONS TO UNSOLVED PROBLEMS

into George's mind, he would be amazed that George saw green skies. However, George has learned to associate the word "blue" with what his mind sees as green, and so he calls the sky "blue", because for him the color green has the name "blue." The question is whether blue must be blue for all people, or whether the <u>perception</u> of that particular color is assigned by the mind.

This extends to all areas of the physical reality, where the outside world we perceive is merely a representation of what is impressed upon the senses. The objects we see are in truth wave-emitting (or reflecting) objects which the brain shows to the conscious self in various forms and colors. Whether the colors and forms experienced perfectly match between person to person, may never be known. That people can communicate accurately shows that the order and proportionality in which experience is interpreted is generally reliable. Thus one's reality is, at least, compatible to another person's in terms of structure and ratio.

SOLUTION:

Qualia: Unusual ability / Usual inability.

...

Nathan Coppedge

SOLUTIONS TO LONG-STANDING PROBLEMS IN PHILOSOPHY

PHILOSOPHY OF MATHEMATICS

PROBLEM: **Mathematical objects**

Main article: Mathematical structure

What are numbers, sets, groups, points, etc.? Are they real objects or are they simply relationships that necessarily exist in all structures? Although many disparate views exist regarding what a mathematical object is, the discussion may be roughly partitioned into two opposing schools of thought: platonism, which asserts that mathematical objects are real, and formalism, which asserts that mathematical objects are merely formal constructions. This dispute may be better understood when considering specific examples, such as the "continuum hypothesis". The continuum hypothesis has been proven independent of the ZF axioms of set theory, so within that system, the proposition can neither be proven true nor proven false. A formalist would therefore say that the continuum hypothesis is neither true nor false, unless you further refine the context of the question. A platonist, however, would assert that there either does or does not exist a transfinite set with a cardinality less than the continuum but greater than any countable set.

So, regardless of whether it has been proven unprovable, the platonist would argue that an answer nonetheless does exist.

THE EQUATIONS AND SOLUTIONS TO UNSOLVED PROBLEMS

SOLUTION:

Numbers not absolutely clear /We can figure or sketch or theorize that it is absolutely undefined, not defined absolutely. Therefore to have a clear picture of numbers we must conflate the definition with the absolute, the definition aims to be complete.

Nathan Coppedge

SOLUTIONS TO LONG-STANDING PROBLEMS IN PHILOSOPHY

PHILOSOPHY OF LANGUAGE:

PROBLEM: Counterfactuals

Main article: [Counterfactual conditional](#)

A counterfactual statement is a conditional statement with a false antecedent. For example, the statement "If [Joseph Swan](#) had not invented the modern [incandescent light bulb](#), then someone else would have invented it anyway" is a counterfactual, because in fact, Joseph Swan invented the modern incandescent light bulb. The most immediate task concerning counterfactuals is that of explaining their truth-conditions. As a start, one might assert that background information is assumed when stating and interpreting counterfactual conditionals and that this background information is just every true statement about the world as it is (pre-counterfactual). In the case of the Swan statement, we have certain trends in the history of technology, the utility of artificial light, the discovery of electricity, and so on. We quickly encounter an error with this initial account: among the true statements will be "Joseph Swan did invent the modern incandescent light bulb." From the conjunction of this statement (call it "S") and the antecedent of the counterfactual ("¬S"), we can derive any conclusion, and we have the unwelcome result that any statement follows from any counterfactual (see the [principle of explosion](#)). [Nelson Goodman](#) takes up this and related issues in his seminal [Fact, Fiction, and Forecast](#); and [David Lewis](#)'s influential articulation of [possible world](#) theory is popularly applied in efforts to solve it.

THE EQUATIONS AND SOLUTIONS TO UNSOLVED PROBLEMS

SOLUTION:

Counterfactuals: Becoming true / Not becoming untrue. True realization is unavoidable by any conditional. The true terms define the conditions without contradiction. Rule of non-contradiction between actual conditionals. Not becoming untrue qua conditional.

SOLUTIONS TO LONG-STANDING PROBLEMS IN PHILOSOPHY

AESTHETICS:

PROBLEM: Essentialism

In art, essentialism is the idea that each <u>medium</u> has its own particular strengths and weaknesses, contingent on its mode of communication. A <u>chase scene</u>, for example, may be appropriate for <u>motion pictures</u>, but poorly realized in <u>poetry</u>, because the essential components of the poetic medium are ill suited to convey the information of a chase scene. This idea may be further refined, and it may be said that the <u>haiku</u> is a poor vehicle for describing a lover's affection, as opposed to the <u>sonnet</u>. Essentialism is attractive to artists, because it not only delineates the role of art and media, but also prescribes a method for evaluating art (quality correlates to the degree of <u>organic form</u>). However, considerable criticism has been leveled at essentialism, which has been unable to formally define organic form or for that matter, medium. What, after all, is the medium of poetry? If it is language, how is this distinct from the medium of prose fiction? Is the distinction really a distinction in medium or <u>genre</u>? Questions about organic form, its definition, and its role in art remain controversial. Generally, working artists accept some form of the concept of organic form, whereas philosophers have tended to regard it as vague and irrelevant.

THE EQUATIONS AND SOLUTIONS TO UNSOLVED PROBLEMS

SOLUTION:

What is art? What is not art has no definition. By a loose standard, art is anything with a definition. Then we know our life is imperfect if definitions are imperfectly realized, since art won't exist if it is not created. Art in the broadest sense is merely a standard of every (each) thing, each it's own art, and a means of fulfilling each standard. For example creating a painting is an art, how to think about a painting is an art, and how to feel about a painting is an art. They cannot be one art without referring to a more general art.

PROBLEM: **Art objects**

This problem originally arose from the practice rather than theory of art. Marcel Duchamp, in the 20th century, challenged conventional notions of what "art" is, placing ordinary objects in galleries to prove that the context rather than content of an art piece determines what art is. In music, John Cage followed up on Duchamp's ideas, asserting that the term "music" applied simply to the sounds heard within a fixed interval of time.

While it is easy to dismiss these assertions, further investigation...

shows that Duchamp and Cage are not so easily disproved. For example, if a pianist plays a Chopin etude, but his finger slips missing one note, is it still the Chopin etude or a new piece of music entirely? Most people would agree that it is still a Chopin etude (albeit with a missing note), which brings into play the Sorites paradox, mentioned below. If one accepts that this is not a fundamentally changed work of music, however, is one implicitly agreeing with Cage that it is merely the duration and context of musical performance, rather than the precise content, which determines what music is? Hence, the question is what the criteria for art objects are and whether these criteria are entirely context-dependent.

SOLUTION: Metaphysical existence of art: It (art or not) might not be metaphysically real/ It is real qua real no matter what. Define real? We have said art merely requires a definition. If you don't define it, that's cheating.

Art answers partly due to Vu, an art professor at Southern CT State U.

SOLUTIONS TO LONG-STANDING PROBLEMS IN PHILOSOPHY

PHILOSOPHY OF SCIENCE

PROBLEM: **Problem of induction**

Intuitively, it seems to be the case that we know certain things with absolute, complete, utter, unshakable certainty. For example, if you travel to the Arctic and touch an iceberg, you know that it would feel cold. These things that we know from experience are known through induction. The problem of induction in short; (1) any inductive statement (like the sun will rise tomorrow) can only be deductively shown if one assumes that nature is uniform. (2) the only way to show that nature is uniform is by using induction. Thus induction cannot be justified deductively.

SOLUTION:

Hume's Arrow / induction: what we don't determine is unknown/ epistemic justification is required. Partly a matter of opinion if we are looking at beliefs.

PROBLEM:

Demarcation problem

Main article: Demarcation problem

'The problem of demarcation' is an expression introduced by Karl Popper to refer to 'the problem of finding a criterion which would enable us to distinguish between the empirical sciences on the one hand, and mathematics and logic as well as "metaphysical" systems on the other'. Popper attributes this problem to Kant. Although Popper mentions mathematics and logic, other writers focus on distinguishing science from metaphysics.

SOLUTION:

Demarcation problem again: Different problems in science are marked by different solutions in metaphysics. Similarities across disciplines are expressed by differences of scope, and similarities across disciplines are expressed by differences of scope. Basically, short-range differences should be viewed in contrast while long-range differences should be viewed for similarity. It's a matter of compare-and-contrast regardless of how opposite the disciplines are, but contrast is the appropriate tool for small differences in oppositeness.

THE EQUATIONS AND SOLUTIONS TO UNSOLVED PROBLEMS

PROBLEM:

Realism Main article: Scientific realism

Does a world independent of human beliefs and representations exist? Is such a world empirically accessible, or would such a world be forever beyond the bounds of human sense and hence unknowable? Can human activity and agency change the objective structure of the world?

SOLUTION:

Realism: The human context might be objective, but we don't know / Where is alienation without a subjective context? Here we find certainty unless we have more confidence in objectivity. Not a matter of knowledge unless the knowledge is real, but apparently everything is what it is no matter what. Knowledge is not special. Reality is a conditional. Everything is as-is, perspective only qualifies perspective. Special knowledge is no more than special knowledge, knowledge not being especially special.

...

G

Nathan Coppedge

eometric Proof of Categorical Deduction

1

When the original lines includes the passing lines, the figure is a form of philosophy.

CUBICAL FIGURES FOR D, IFS(D=1,"3 ORIGINAL LINES, 1 PASSING LINE, AND 1 LINE BROKEN ONCE",D>1,(3*(2^(D-1))&" ORIGINAL LINES, "&(2^(D-1))&" PASSING LINES, AND "&(2^(D-1))&" LINES BROKEN "&(2^(D-1))&" TIMES")) If the number of original lines includes passing lines, then the figure may include categoric opposites opposed diagonally. —Simple Infinite Geometry

2

ANOTHER GEOMETRY PROOF: D = 1, Area 2 is now at the base of the simple shapes, and has a predicted area of '1'. This is a surprising result. This may predict that D = 0 has a natural base area of 1. This may predict that unity is standard for the zeroth dimension indicating a coherent unity. —Perimeter of Hypercubes Calculator

SOLUTIONS TO UN-SOLVED PROBLEMS IN PHYSICS

[Problems are taken from: Wikipedia Unsolved Problems in Physics]

General Physics / Quantum Physics

PROBLEM: Arrow of time (e.g. entropy's arrow of time): Why does time have a direction? Why did the universe have such low entropy in the past, and time correlates with the universal (but not local) increase in entropy, from the past and to the future, according to the second law of thermodynamics? [4] Why are CP violations observed in certain weak force decays, but not elsewhere? Are CP violations somehow a product of the Second Law of Thermodynamics, or are they a separate arrow of time? Are there exceptions to the principle of causality? Is there a single possible past? Is the present moment physically distinct from the past and future, or is it merely an emergent property of consciousness? What links the quantum arrow of time to the thermodynamic arrow?

SOLUTION:

Asymmetric Time / Some Type of Parallelism Creating Group Behavior, for Example, in Consciousness and Wormholes.

PROBLEM:

Interpretation of quantum mechanics: How does the quantum description of reality, which includes elements such as the superposition of states and wavefunction collapse or quantum decoherence, give rise to the reality we perceive?[4] Another way of stating this question regards the measurement problem: What constitutes a "measurement" which apparently causes the wave function to collapse into a definite state? Unlike classical physical processes, some quantum mechanical processes (such as quantum teleportation arising from quantum entanglement) cannot be simultaneously "local", "causal", and "real", but it is not obvious which of these properties must be sacrificed,[5] or if an attempt to describe quantum mechanical processes in these senses is a category error such that a proper understanding of quantum mechanics would render the question meaningless.

SOLUTION:

Paradoxical Observation / Chirality / Flavor/ Differentiation of Nature, Non-Quantum (Non-X) Continuum, Rule of Dimensional Variation and Perhaps Non-Dimensional Nature.

THE EQUATIONS AND SOLUTIONS TO UNSOLVED PROBLEMS

PROBLEM: Grand Unification Theory/ Theory of everything: Is there a theory which explains the values of all fundamental physical constants?[4] Is there a theory which explains why the gauge groups of the standard model are as they are, and why observed spacetime has 3 spatial dimensions and 1 temporal dimension? Do "fundamental physical constants" vary over time? Are any of the fundamental particles in the standard model of particle physics actually composite particles too tightly bound to observe as such at current experimental energies? Are there fundamental particles that have not yet been observed, and, if so, which ones are they and what are their properties? Are there unobserved fundamental forces?

SOLUTION:

Specific (Gravity, etc)/ Coherent Assembly. Emergent problem-solving as mind-over-matter rule.

PROBLEM:

Yang–Mills theory: Given an arbitrary compact gauge group, does a non-trivial quantum Yang–Mills theory with a finite mass gap exist? This problem is also listed as one of the Millennium Prize Problems in mathematics.[6]

SOLUTION:

Dimensional beautiful layers problem / Equal conjoint solution, compact, broken or else strange.

PROBLEM: Physical information: Are there physical phenomena, such as wave function collapse or black holes, which irrevocably destroy information about their prior states? [7] How is quantum information stored as a state of a quantum system?

SOLUTION:

Bulk rarity likilihood (conjecture) / garbage in garbage out, survival of rare common (common rarity) garbage theory, survival of universal singularities == rare common garbage.

THE EQUATIONS AND SOLUTIONS TO UNSOLVED PROBLEMS

PROBLEM: <u>Dimensionless physical constant</u>: At the present time, the values of the dimensionless physical constants cannot be calculated; they are determined only by physical measurement.[8][9] What is the minimum number of dimensionless physical constants from which all other dimensionless physical constants can be derived? Are dimensional physical constants necessary at all?

SOLUTION:

Parsimony of physical quanta / Finite unlimited = Infinity, all true zero-dimensional constants are infinite.

PROBLEM: <u>Fine-tuned Universe</u>: The values of the fundamental physical constants are in a narrow range necessary to support carbon-based life.[10][11][12] Is this because there exist <u>other universes</u> with different constants, or are our universe's constants the result of chance, or some other factor or process?

SOLUTION:

Our complex paradigmatic function / Simple choices. Choice paradigm. Choices are paradigmatic. (Further clarification: Paradigmatic survival. Individuation, evolution of the universe. Part of universe paradigm. Human significance is non-negative equated with universal empiricism).

189

PROBLEM: Quantum field theory: Is it possible to construct, in the mathematically rigorous framework of algebraic QFT, a theory in 4-dimensional spacetime that includes interactions and does not resort to perturbative methods?[13][14]

SOLUTION:

Refined quantum timeless / See first problem, sort of the opposite of that: unparallel time.

THE EQUATIONS AND SOLUTIONS TO UNSOLVED PROBLEMS

SOLUTIONS TO LONG-STANDING PROBLEMS IN PHYSICS

Cosmology and GR

PROBLEM: Problem of time: In quantum mechanics time is a classical background parameter and the flow of time is universal and absolute. In general relativity time is one component of four-dimensional spacetime, and the flow of time changes depending on the curvature of spacetime and the spacetime trajectory of the observer. How can these two concepts of time be reconciled?[15]

SOLUTION:

Ambiguous time constants / Arbitrary universal quantities possibly involving wormholes. Arbitrary duration continuum with physical properties.

PROBLEM: <u>Cosmic inflation</u>: Is the theory of cosmic inflation in the very early universe correct, and, if so, what are the details of this epoch? What is the hypothet-ical <u>inflatonscalar field</u> that gave rise to this cosmic inflation? If inflation happened at one point, is it <u>self-sustaining through inflation of quantum-mechanical fluctuations</u>, and thus ongoing in some extremely distant place?[16]

SOLUTION: Big bang / creation from noth-ingness (expansion problem, etc) / Specific objects die, the overall picture is unchanging. Expansion doesn't make sense because of the size of the universe. The visible universe is much smaller, perhaps infinitely smaller than the actual universe, and there's no telling how much empty space might seperate real things, not to mention virtual things. Perhaps virtual space seperares real things and absolute space seperates virtual things.

THE EQUATIONS AND SOLUTIONS TO UNSOLVED PROBLEMS

PROBLEM: Horizon problem: Why is the distant universe so homogeneous when the Big Bang theory seems to predict larger measurable anisotropies of the night sky than those observed? Cosmological inflation is generally accepted as the solution, but are other possible explanations such as a variable speed of light more appropriate?[17]

SOLUTION:

Extrapolating non-uniformity from constants / Non-constants predict uniformity.

PROBLEM: Origin and future of the universe: How did the conditions for anything to exist arise? Is the universe heading towards a Big Freeze, a Big Rip, a Big Crunch, or a Big Bounce? Or is it part of an infinitely recurring cyclic model?

SOLUTION:

Question of the consistency of everything / inconsistency of nothing, everything by it's own rule.

PROBLEM: Size of universe: The diameter of the observable universe is about 93 billion light-years, but what is the size of the whole universe? Does a multiverse exist?

SOLUTION:

Local everything 93 billion lightyears / Small matter, 1/93 universal nothings.

PROBLEM: Baryon asymmetry: Why is there far more matter than antimatter in the observable universe?

SOLUTION:

Antimatter matters / Matter is against it.

PROBLEM: Cosmological constant problem: Why does the zero-point energy of the vacuum not cause a large cosmological constant? What cancels it out?[18][19]

SOLUTION:

Zero resistance justified by infinite opposite resistance.

THE EQUATIONS AND SOLUTIONS TO UNSOLVED PROBLEMS

PROBLEM: Dark matter: What is the identity of dark matter?[17] Is it a particle? Is it the lightest superpartner (LSP)? Or, do the phenomena attributed to dark matter point not to some form of matter but actually to an extension of gravity?

SOLUTION:

Mysterious matter / Obvious antimatter, antimatter manifestations.

PROBLEM: Dark energy: What is the cause of the observed accelerated expansion (de Sitter phase) of the universe? Why is the energy density of the dark energy component of the same magnitude as the density of matter at present when the two evolve quite differently over time; could it be simply that we are observing at exactly the right time? Is dark energy a pure cosmological constant or are models of quintessence such as phantom energy applicable?

SOLUTION:

Mystery energy expansion /Obvious collapsed entropy, perhaps minimum energy threshold.

PROBLEM: <u>Dark flow</u>: Is a non-spherically symmetric gravitational pull from outside the observable universe responsible for some of the observed motion of large objects such as galactic clusters in the universe?

SOLUTION:

Big scary mass tractor beam theory / Little bit of antigravity effect.

PROBLEM: <u>Axis of evil</u>: Some large features of the microwave sky at distances of over 13 billion light years appear to be aligned with both the motion and orientation of the solar system. Is this due to systematic errors in processing, contamination of results by local effects, or an unexplained violation of the <u>Copernican principle</u>?

SOLUTION:

Macro effects too quantum / meso / micro not quantum enough.

THE EQUATIONS AND SOLUTIONS TO UNSOLVED PROBLEMS

PROBLEM: Shape of the universe: What is the 3-manifold of comoving space, i.e. of a comoving spatial section of the universe, informally called the "shape" of the universe? Neither the curvature nor the topology is presently known, though the curvature is known to be "close" to zero on observable scales. The cosmic inflation hypothesis suggests that the shape of the universe may be unmeasurable, but, since 2003, Jean-Pierre Luminet, et al., and other groups have suggested that the shape of the universe may be the Poincaré dodecahedral space. Is the shape unmeasurable; the Poincaré space; or another 3-manifold?

SOLUTION:

Scalar shape / Scalelessly shapeless.

Nathan Coppedge

SOLUTIONS TO LONG-STANDING PROBLEMS IN PHYSICS

QUANTUM GRAVITY PROBLEMS

PROBLEM: Vacuum catastrophe: Why does the predicted mass of the quantum vacuum have little effect on the expansion of the universe?[19]

SOLUTION:

Time constant predicts separate space.

PROBLEM: Quantum gravity: Can quantum mechanics and general relativity be realized as a fully consistent theory (perhaps as a quantum field theory)?[20] Is spacetime fundamentally continuous or discrete? Would a consistent theory involve a force mediated by a hypothetical graviton, or be a product of a discrete structure of spacetime itself (as in loop quantum gravity)? Are there deviations from the predictions of general relativity at very small or very large scales or in other extreme circumstances that flow from a quantum gravity theory?

SOLUTION:

Quantum gravity: arbitrary energy destroys itself.

PROBLEM: Black holes, black hole information paradox, and black hole radiation: Do black holes produce thermal radiation, as expected on theoretical grounds?[7] Does this radiation contain information about their inner structure, as suggested by gauge–gravity duality, or not, as implied by Hawking's original calculation? If not, and black holes can evaporate away, what happens to the information stored in them (since quantum mechanics does not provide for the destruction of information)? Or does the radiation stop at some point leaving black hole remnants? Is there another way to probe their internal structure somehow, if such a structure even exists?

SOLUTION:

Information ecliptic singularity / What escapes the universe or universal has no information, regardless of interaction with black holes. Wormholes may be implicated.

Nathan Coppedge

PROBLEM: <u>Extra dimensions</u>: Does nature have more than four <u>spacetime</u>dimensions? If so, what is their size? Are dimensions a fundamental property of the universe or an emergent result of other physical laws? Can we experimentally observe evidence of higher spatial dimensions?

SOLUTION:

What is the wormhole of the universe? The mind. What is the mind of the universe? A wormhole. What is the wormhole of tjhe mind? Universal time.

THE EQUATIONS AND SOLUTIONS TO UNSOLVED PROBLEMS

PROBLEM: The cosmic censorship hypothesis and the chronology protection conjecture: Can singularities not hidden behind an event horizon, known as "naked singularities", arise from realistic initial conditions, or is it possible to prove some version of the "cosmic censorship hypothesis" of Roger Penrose which proposes that this is impossible?
[21] Similarly, will the closed timelike curves which arise in some solutions to the equations of general relativity (and which imply the possibility of backwards time travel) be ruled out by a theory of quantum gravity which unites general relativity with quantum mechanics, as suggested by the "chronology protection conjecture" of Stephen Hawking?

SOLUTION:

Way-out from destruction (conjecture) / Inescapable options scale to theory. There is a theoretical basis for time-travel, which is sufficient.

Nathan Coppedge

PROBLEM:

Locality: Are there non-local phenomena in quantum physics?[22][23] If they exist, are non-local phenomena limited to the entanglement revealed in the violations of the Bell inequalities, or can information and conserved quantities also move in a non-local way? Under what circumstances are non-local phenomena observed? What does the existence or absence of non-local phenomena imply about the fundamental structure of spacetime? How does this elucidate the proper interpretation of the fundamental nature of quantum physics?

SOLUTION:

Possibility of an association solution / Impossibility of a non-association problem.

SOLUTIONS TO LONG-STANDING PROBLEMS IN PHYSICS

HIGH-ENERGY PHYSICS / PARTICLE PHYSICS PROBLEMS

PROBLEM: Hierarchy problem: Why is gravity such a weak force? It becomes strong for particles only at the Planck scale, around 10^{19} GeV, much above the electroweak scale (100 GeV, the energy scale dominating physics at low energies). Why are these scales so different from each other? What prevents quantities at the electroweak scale, such as the Higgs boson mass, from getting quantum corrections on the order of the Planck scale? Is the solution supersymmetry, extra dimensions, or just anthropic fine-tuning?

SOLUTION:

Weak gravity / Strong antigravity.

PROBLEM: <u>Planck particle</u>: The Planck mass plays an important role in parts of mathematical physics. A series of researchers have suggested the existence of a fundamental particle with mass equal to or close to that of the Planck mass. The Planck mass is however enormous compared to any detected particle even compared to the Higgs particle. It is still an unsolved problem if there exist or even have existed a particle with close to the Planck mass. This is indirectly related to the hierarchy problem.

SOLUTION:

Planck particles: Realizing mass limits / Characterized by no possibility of massless infinity.

THE EQUATIONS AND SOLUTIONS TO UNSOLVED PROBLEMS

PROBLEM: <u>Magnetic monopoles</u>: Did particles that carry "magnetic charge" exist in some past, higher-energy epoch? If so, do any remain today? (<u>Paul Dirac</u>showed the existence of some types of magnetic monopoles would explain <u>charge quantization</u>.) [24]

SOLUTION:

Magnetic monopoles: Timeless quality question / Types of time answer.

PROBLEM: <u>Proton decay</u> and <u>spin crisis</u>: Is the proton fundamentally stable? Or does it decay with a finite lifetime as predicted by some extensions to the standard model? [25] How do the quarks and gluons carry the spin of protons?[26]

SOLUTION:

Neutral nuclear prediction / Polarized denuclearized memory, perhaps electrons record / transcribe neutrons.

PROBLEM: <u>Supersymmetry</u>: Is spacetime supersymmetry realized at TeV scale? If so, what is the mechanism of supersymmetry breaking? Does supersymmetry stabilize the electroweak scale, preventing high quantum corrections? Does the lightest<u>supersymmetric particle</u> (LSP or <u>Lightest Supersymmetric Particle</u>) comprise <u>dark matter</u>?

SOLUTION:

Macroscopic symmetry question / micro asymmetry answer. For example, if an entire galaxy is symmetric, there might be small differences. If a large particle is symmetric, this might produce inconsistencies due to high energy and sectional asymmetry.

PROBLEM: <u>Generations of matter</u>: Why are there three generations of <u>quarks</u> and <u>leptons</u>? Is there a theory that can explain the masses of particular quarks and leptons in particular generations from first principles (a theory of<u>Yukawa couplings</u>)?[27]

SOLUTION: Inconsistent symmetric coupling / Consistent decoherence with asymmetry. E.g. occasional 4-d coupling increases energy.

THE EQUATIONS AND SOLUTIONS TO UNSOLVED PROBLEMS

PROBLEM: Neutrino mass: What is the mass of neutrinos, whether they follow Dirac or Majorana statistics? Is the mass hierarchy normal or inverted? Is the CP violating phase equal to 0?[28][29]

SOLUTION: Neutrinos: Uncertain low-mass ejection / Certain high-mass retained. Neutrinos are gravity-related particles. They might derive their mass from gravity.

PROBLEM: Colour confinement: Why has there never been measured a free quark or gluon, but only objects that are built out of them, such as mesons and baryons? How does this phenomenon emerge from QCD?

SOLUTION: Transient mass flavors (gluons, etc) / Flavorless opposition in the form of permanence. Permanence influence.

PROBLEM: Strong CP problem and axions: Why is the strong nuclear interaction invariant to parity and charge conjugation? Is Peccei–Quinn theory the solution to this problem? Could axions be the main component of dark matter?

SOLUTION: Dipolar categories / Universal monopole.

PROBLEM: <u>Anomalous magnetic dipole moment</u>: Why is the experimentally measured value of the <u>muon</u>'s anomalous magnetic dipole moment ("muon g–2") significantly different from the theoretically predicted value of that physical constant?[30]

SOLUTION: Bizarre magnetic ecliptic: Normalized dipole repulsion. Perhaps unknown smaller particle or influence such as asymmetry, ovular shape, interacrion with neutrinos, or long-distance interactions. I am not clear on the problem. Perhaps shared electrons? Mismeasurement? Higher-dimensional or lower-dimensional properties?

PROBLEM: <u>Proton radius puzzle</u>: What is the electric <u>charge radius</u> of the proton? How does it differ from gluonic charge?

SOLUTION:

Permanent charge boundary / Dynamic neutral interaction. Perhaps the muon is evenly distrubuted due to interaction with neutrons.

THE EQUATIONS AND SOLUTIONS TO UNSOLVED PROBLEMS

PROB LEM: <u>Pentaquarks</u> and other <u>exotic hadrons</u>: What combinations of quarks are possible? Why were pentaquarks so difficult to discover?[31] Are they a tightly-bound system of five elementary particles, or a more weakly-bound pairing of a baryon and a meson?[32]

SOLUTION:

Strange particles / Typical exotics.

PROBLEM: <u>Mu problem</u>: problem of <u>supersymmetric</u>theories, concerned with understanding the parameters of the theory.

SOLUTION:

Mu problem: Free properties/confined quanta.

Nathan Coppedge

PROBLEM: <u>Koide formula</u>: An aspect of the <u>problem of particle generations</u>. The sum of the masses of the three charged leptons, divided by the square of the sum of the roots of these masses is $Q=23Q=23$, to within one standard deviation of observations. It is unknown how such a simple value comes about, and why it is the exact arithmetic average of the possible extreme values of 1/3 (equal masses) and 1 (one mass dominates).

SOLUTION:

Antipolar spin / Axis of spin, for example common axes or ecliptic relations.

THE EQUATIONS AND SOLUTIONS TO UNSOLVED PROBLEMS

SOLUTIONS TO LONG-STANDING PROBLEMS IN PHYSICS

ASTRONOMY AND ASTROPHYSICS PROBLEMS

PROBLEM: Astrophysical jet: Why do only certain accretion discs surrounding certain astronomical objects emit relativistic jets along their polar axes? Why are there quasi-periodic oscillations in many accretion discs?[33] Why does the period of these oscillations scale as the inverse of the mass of the central object?[34] Why are there sometimes overtones, and why do these appear at different frequency ratios in different objects?[35]

SOLUTION:

Black Hole Jets: Mysterious high-mass repulsed heat/Ordinary low mass low temperature attraction, relativistic hot ice.

PROBLEM: Diffuse interstellar bands: What is responsible for the numerous interstellar absorption lines detected in astronomical spectra? Are they molecular in origin, and if so which molecules are responsible for them? How do they form?

SOLUTION:

Interstellar hot dust composition / Stars decompose, cool, sometimes not dust.

PROBLEM: Supermassive black holes: What is the origin of the M-sigma relation between supermassive black hole mass and galaxy velocity dispersion?[36] How did the most distant quasars grow their supermassive black holes up to 1010 solar masses so early in the history of the universe?

SOLUTION:

Local giants / Non-locally tiny. Universe is bigger than we think, distant supermassive black holes are remnants of a relation to further parts of the universe, perhaps supermassive galaxies or supermassive star clusters.

THE EQUATIONS AND SOLUTIONS TO UNSOLVED PROBLEMS

PROBLEM: Kuiper cliff: Why does the number of objects in the Solar System's Kuiper belt fall off rapidly and unexpectedly beyond a radius of 50 astronomical units?

SOLUTION:

Massive spiral speed observed, this means there is not accretion of masses, or black holes have an alternate source of energy.

PROBLEM:

Flyby anomaly: Why is the observed energy of satellites flying by Earth sometimes different by a minute amount from the value predicted by theory?

SOLUTION:

Asteroids 50 astronomical units problem / The solution involves 1/50 asteroids obliterated out of range of sun. Perhaps long age of universe or more active period, or unknown phenomena such as asteroid mining.

PROBLEM:

Galaxy rotation problem: Is dark matterresponsible for differences in observed and theoretical speed of stars revolving around the centre of galaxies, or is it something else?

SOLUTION:

Unlikely accuracy → Likely probability. Instruments are using more probability than you think to make their estimates. Perhaps improving formulas particularly for small objects cost too much money. Plus, some objects are affected by Earth's atmosphere, decomposition, the heat of the sun, scientific tampering, quantum effects, technology, solar wind, etc. Also, gravitational waves.

THE EQUATIONS AND SOLUTIONS TO UNSOLVED PROBLEMS

PROBLEM: <u>Supernovae</u>: What is the exact mechanism by which an implosion of a dying star becomes an explosion?

SOLUTION:

Non-elastic center / electrostatic, electric expansion / contraction of velocity with high energy close masses, perhaps related to black holes. Static electricity.

PROBLEM:

<u>p-nuclei</u>: What astrophysical process is responsible for the <u>nucleogenesis</u> of these rare isotopes?

SOLUTION:

P-nuclei, as far as I can tell explosive metals. Basically created by massive amounts of metal undergoing fusion.

PROBLEM:

Ultra-high-energy cosmic ray:[17] Why is it that some cosmic rays appear to possess energies that are impossibly high, given that there are no sufficiently energetic cosmic ray sources near the Earth? Why is it that (apparently) some cosmic rays emitted by distant sources have energies above the Greisen–Zatsepin–Kuzmin limit?[4][17]

SOLUTION:

Contraction of mass / Release of energy from heat and gravity, chain reaction, not a problem if energy is lost. Chemical reaction with bosons or gravitons and extremely, perhaps higher-than-predicted, high-energy states.

THE EQUATIONS AND SOLUTIONS TO UNSOLVED PROBLEMS

PROBLEM: Rotation rate of <u>Saturn</u>: Why does the <u>magnetosphere of Saturn</u> exhibit a (slowly changing) periodicity close to that at which the planet's clouds rotate? What is the true rotation rate of Saturn's deep interior? [37]

SOLUTION:

High-energy gamma rays: Cannot penetrate the impossible / Possibility it broke a barrier.

PROBLEM: Origin of <u>magnetar magnetic field</u>: What is the origin of <u>magnetar</u> magnetic field?

SOLUTION:

Non-coincidental magnetic reaction / Coincidental non-magnetic cause. Probably a coincidence with the chemical reactions.

PROBLEM: Large-scale anisotropy: Is the universe at very large scales anisotropic, making the cosmological principle an invalid assumption? The number count and intensity dipole anisotropy in radio, NRAO VLA Sky Survey (NVSS) catalogue[38] is inconsistent with the local motion as derived from cosmic microwave background[39][40] and indicate an intrinsic dipole anisotropy. The same NVSS radio data also shows an intrinsic dipole in polarization density and degree of polarization[41] in the same direction as in number count and intensity. There are other several observation revealing large-scale anisotropy. The optical polarization from quasars shows polarization alignment over a very large scale of Gpc.[42][43][44] The cosmic-microwave-background data shows several features of anisotropy,[45][46][47][48]which are not consistent with the Big Bang model.

SOLUTION: Moderately expansive magnetic field / Hyper-condensed gravity waves.

THE EQUATIONS AND SOLUTIONS TO UNSOLVED PROBLEMS

PROBLEM: Space roar: Why is space roar six times louder than expected? What is the source of space roar?

SOLUTION:

Distant noise grows near / Nearby quiet shrinks. This may have to do with the 0-dimensionality of space.

PROBLEM: Age–metallicity relation in the Galactic disk: Is there a universal age–metallicity relation (AMR) in the Galactic disk (both "thin" and "thick" parts of the disk)? Although in the local (primarily thin) disk of the Milky Waythere is no evidence of a strong AMR,[49] a sample of 229 nearby "thick" disk stars has been used to investigate the existence of an age–metallicity relation in the Galactic thick disk, and indicate that there is an age–metallicity relation present in the thick disk.[50][51] Stellar ages from asteroseismology confirm the lack of any strong age-metallicity relation in the Galactic disc.[52]

SOLUTION: Symmetricity problem involving polarization of the whole observable universe / Solution is non-observable asymmetric singularity in part of the universe--e.g. observed from a non-symmetric location.

PROBLE M:

The lithium problem: Why is there a discrepancy between the amount of lithium-7 predicted to be produced in Big Bang nucleosynthesis and the amount observed in very old stars?[53]

SOLUTION:

Relatively massive and relatively energetic is exaggerated over long time periods.

PROBLEM:

Ultraluminous pulsar: The ultraluminous X-ray source M82 X-2 was thought to be a black hole, but in October 2014 data from NASA's space-based X-ray telescope NuStarindicated that M82 X-2 is a pulsar many times brighter than the Eddington limit.

SOLUTION:

Super-bright pulsar: any mere number can be exceeded by raising the reality quotient.

THE EQUATIONS AND SOLUTIONS TO UNSOLVED PROBLEMS

PROBLEM: <u>Fast radio bursts</u>: Transient radio pulses lasting only a few milliseconds, from emission regions thought to be no larger than a few hundred kilometres, and estimated to occur several hundred times a day. While several theories have been proposed, there is no generally accepted explanation for them. The only known *repeating* FRB emanates from a galaxy roughly 3 billion light years from Earth.[54][55]

SOLUTION:

Don't want to lose the signal / Save our star.

Nathan Coppedge

SOLUTIONS TO LONG-STAN DING PROBLEMS IN PHYSICS

NUCLEAR PHYSICS PROBLEMS

PROBLEM: Quantum chromodynamics: What are the phases of strongly interacting matter, and what roles do they play in the evolution of cosmos? What is the detailed partonic structure of the nucleons? What does QCD predict for the properties of strongly interacting matter? What determines the key features of QCD, and what is their relation to the nature of gravity and spacetime? Do glueballs exist? Do gluons acquire mass dynamically despite having a zero rest mass, within hadrons? Does QCD truly lack CP-violations? Do gluons saturate when their occupation number is large? Do gluons form a dense system called Colour Glass Condensate? What are the signatures and evidences for the Balitsky-Fadin-Kuarev-Lipatov, Balitsky-Kovchegov, Catani-Ciafaloni-Fiorani-Marchesini evolution equations?

SOLUTION

Strong energy /Translucent structure.

THE EQUATIONS AND SOLUTIONS TO UNSOLVED PROBLEMS

PROBLEM: <u>Nuclei</u> and <u>nuclear astrophysics</u>: What is the nature of the <u>nuclear force</u> that binds <u>protons</u> and <u>neutrons</u> into <u>stable nuclei</u> and rare isotopes? What is the nature of exotic excitations in nuclei at the frontiers of stability and their role in stellar processes? What is the nature of <u>neutron stars</u> and dense <u>nuclear matter</u>? What is the origin of the elements in the <u>cosmos</u>? What are the nuclear reactions that drive <u>stars</u> and stellar explosions?

SOLUTION:

Inverse flavor chemistry with heavy compounfds, for example molten sulfur is conducive to water.

SOLUTIONS TO LONG-STANDING PROBLEMS IN PHYSICS

ATOMIC, MOLECULAR, AND OPTICAL PHYSICS PROBLEMS

PROBLEM: Abraham–Minkowski controversy: What is the momentum of light in optical media?

SOLUTION:

Abraham-Minkowsky Controversy: Quantum 3-reactions predict perpetual motion and are not against nuclear chemistry.

PROBLEM:

Bose–Einstein condensation: How do we rigorously prove the existence of Bose–Einstein condensates for general interacting systems? [56]

SOLUTION:

Bose-Einstein Condensates: Infinite magnetism in general cases/ Ecliptic in a gravity well with exotic matter.

SOLUTIONS TO LONG-STANDING PROBLEMS IN PHYSICS

CLASSICAL MECHANICS PROBLEMS

PROBLEM: Singular trajectories in the Newtonian N-body problem: Does the set of initial conditions for which particles that undergo near-collisions gain infinite speed in finite time have measure zero? This is known to be the case when $N \leq 4$, but the question remains open for larger N.[57][58]

SOLUTION:

Sudden jolt / Gradual stop. Doesn't matter if you gain infinite speed, as it will be gone almost immediately (due to Scarpa).

SOLUTIONS TO LONG-STANDING PROBLEMS IN PHYSICS

CONDENSED MATTER PHYSICS PROBLEMS

PROBLEM:

High-temperature superconductors: What is the mechanism that causes certain materials to exhibit superconductivity at temperatures much higher than around 25kelvin? Is it possible to make a material that is a superconductor at room temperature?[4]

SOLUTION:

Full temperature electricity conduction / Some heat involved in conduction.

PROBLEM:

Amorphous solids: What is the nature of the glass transition between a fluid or regular solid and a glassy phase? What are the physical processes giving rise to the general properties of glasses and the glass transition?[59][60]

SOLUTION:

Heating of glass (silicate) creates susceptibility to gravity, a liquid state. Cooling creates fracturing similar to sand. Heating creates the outer texture, the inner composition when cooled is more similar to sand or magma. Some glasses are more perfect for human use than others.

PROBLEM:

Cryogenic electron emission: Why does the electron emission in the absence of light increase as the temperature of a photomultiplier is decreased?[61][62]

SOLUTION:

Obvious electron in the dark / Mysterious anti-electron photon combination.

PROBLEM:

Sonoluminescence: What causes the emission of short bursts of light from imploding bubbles in a liquid when excited by sound?[63][64]

SOLUTION:

Collapse, light not produced by bubble / expansion (of light) produced by dark outside bubble collapse.

PROBLEM:

Turbulence: Is it possible to make a theoretical model to describe the statistics of a turbulent flow (in particular, its internal structures)?[4]Also, under what conditions do smooth solutions to the Navier–Stokes equations exist? The latter problem is also listed as one of the Millennium Prize Problems in mathematics.

SOLUTION:

Navier-Stokes or maybe Black-Scholles: Non-arbitrary extra-extenuated limit / Arbitrary when infinitely divided.

THE EQUATIONS AND SOLUTIONS TO UNSOLVED PROBLEMS

PROBLEM: <u>Alfvénic turbulence</u>: In the solar wind and the turbulence in solar flares, coronal mass ejections, and magnetospheric substorms are major unsolved problems in space plasma physics.[65]

SOLUTION:

Magnetospheric ejections: chemical retensions.

PROBLEM:

<u>Topological order</u>: Is topological order stable at non-zero <u>temperature</u>? Equivalently, is it possible to have three-dimensional <u>self-correctingquantum memory</u>?[66]

SOLUTION

Identity with temporal isomorphism / Disintegral with 4-d properties. Time / universe, consciousness, wormole model.

PROBLEM:

Fractional Hall effect: What mechanism explains the existence of the u=5/2u=5/2state in the fractional quantum Hall effect? Does it describe quasiparticles with non-Abelian fractional statistics?[67]

SOLUTION:

Implicarions of fractional lensing / There is no result from full valences (?)

PROBLEM:

Liquid crystals: Can the nematic to smectic (A) phase transition in liquid crystal states be characterized as a universal phase transition? [68][69]

SOLUTION:

Liquid crystals 'A' phase expressed by: Period, followed by? / Amplitude then transition. The state changes before the phase.

PROBLEM: Semiconductor nanocrystals: What is the cause of the nonparabolicity of the energy-size dependence for the lowest optical absorption transition of quantum dots?[70]

SOLUTION:

Heterogenous rejection / Homogeneous absorption. For example, with dark materials or low light.

PROBLEM:

Metal whiskering: In electrical devices, some metallic surfaces may spontaneously grow fine metallic whiskers, which can lead to equipment failures. While compressive mechanical stress is known to encourage whisker formation, the growth mechanism has yet to be determined.

Metal whiskering / Electromagnetism.

Nathan Coppedge

SOLUTIONS TO LONG-STANDING PROBLEMS IN PHYSICS

PLASMA PHYSICS PROBLEMS

PROBLEM:

Plasma physics and fusion power: Fusion energy may potentially provide power from abundant resource (e.g. hydrogen) without the type of radioactive waste that fission energy currently produces. However, can ionized gases (plasma) be confined long enough and at a high enough temperature to create fusion power? What is the physical origin of H-mode?[71]

SOLUTION:

Atomic waste / Gravity or antimatter power.

THE EQUATIONS AND SOLUTIONS TO UNSOLVED PROBLEMS

PROBLEM:

Solar cycle: How does the Sun generate its periodically reversing large-scale magnetic field? How do other solar-like stars generate their magnetic fields, and what are the similarities and differences between stellar activity cycles and that of the Sun?[72] What caused the Maunder Minimum and other grand minima, and how does the solar cycle recover from a minima state?

SOLUTION:

Sustained magnetism / Unsustainable gravity. Cause of depressurization is chemical pressure.

PROBLEM: Coronal heating problem: Why is the Sun's corona (atmosphere layer) so much hotter than the Sun's surface? Why is the magnetic reconnection effect many orders of magnitude faster than predicted by standard models?

SOLUTION:

Massive gravity doesn't burn easily. / Lightweight particles burn easily.

PROBLEM:

The injection problem: Fermi acceleration is thought to be the primary mechanism that accelerates astrophysical particles to high energy. However, it is unclear what mechanism causes those particles to initially have energies high enough for Fermi acceleration to work on them.[73]

SOLUTION:

Limited by gravity, we would expect deceleration, caused by energy loss/ Escaping particles are accelerated by a heat reaction, e.g. like hot gas. Many reactions are lightweight by cosmological standards.

PROBLEM:

Solar wind interaction with comets: In 2007 the *Ulysses* spacecraft passed through the tail of comet C/2006 P1 (McNaught) and found surprising results concerning the interaction of the solar wind with the tail.

SOLUTION:

Light and heat / Shadow felt cold.

SOLUTIONS TO LONG-STANDING PROBLEMS IN PHYSICS

BIOPHYSICS PROBLEMS

PROBLEM: Stochasticity and robustness to noise in gene expression: How do genes govern our body, withstanding different external pressures and internal stochasticity? Certain models exist for genetic processes, but we are far from understanding the whole picture, in particular in development where gene expression must be tightly regulated.

SOLUTION: Biological Robustness: Unexpected development / Likely not developed. Vis a vis thresholds of tolerance—sensitivity might imply sophisticated brain.

PROBLEM: Quantitative study of the immune system: What are the quantitative properties of immune responses? What are the basic building blocks of immune system networks?

SOLUTION:

Immune systems: Quality of response / Quantity of effects.

PROBLEM:

Unified brain processing theory: How to unify physics and neuroscience?[74]

SOLUTION:

Physical brain identity process/ virtually basic disintegral state or property.

PROBLEM: Homochirality: What is the origin of the preponderance of specific enantiomers in biochemical systems?

SOLUTION:

The origin of amino acids and other parts in humans exhibiting handededness derives from the history of chemical processes suited to survival. Asymmetric code☐ symmetric reading. Symmetric readings are useful for 3-d processes, particularly with organic shapes.

...

SOLUTIONS TO LONGSTANDING PROBLEMS IN SOCIETY

(List doesn't even exist on wikipedia so I, 88% introvert, have founded Social Theory on Quora at 3:30am on 2022–11–08)

- Solution to the Mega-Projects Problem: (1) Over time: Most perpetual motion projects can even be mass-manufactured if done correctly. This cuts down on time. (2) Over budget: Perpetual motion tends to be worth money when it works, so that means that it is not over-budget if manufactured using basic or recycled materials. (3) Over and over again: The math and construction of perpetual motion tends to be relatively easy to replicate once a pattern is set where the right properties are followed. Thus, it is not a problem 'over and over again' anymore.
- Mortality, possibly solved by touching a long wooden pole to the ceiling.
- Torture: if people believe the meaningful life is better, this could prevent torture provided that torture is not meaningful.
- The solution to fear is possibly to have greatness or a group of friends.

The individual path is the immortal path, the societal path is the enlightened path. [Society brings death to the individual, the individual brings death to society].—Politics Links (...)

- Social Metaphysics. Is this true? I don't know. Maybe it is something to solve, or maybe it is not. (1) It might be best to be a woman on the outside. Who wants hell really? It might be someone suing someone or something. All men want is the good life. They're just trying to be inside a woman. They already believe that women are metaphysics. The male soul must be nothing, the female soul must be everything other than her man. Sex is correspondence theory, that is a nice way to put it. (2) The further you go in a single direction, the more you will achieve.
- The Greatness Problem or Limited Cheating Principle: I have thought that no person in history except perhaps [a divine being] has achieved more than 2.5 / 3 things, those things being Extreme Wealth, Unquestioned Pop-

ularity, and Utter Originality. Humans rarely if ever achieve more than two out of the three properties of Wealth, Popularity, and Originality. Achieving 2.5 properties has happened in a small number of cases which involved cheating: for example, Jesus became famous and obscure, Henry Ford became a poor rich boy, and Nathan Coppedge became a stupid genius.

The Greatness Problem for Women: A perfect man, like a perfect woman, will very likely not have more than 2.5 points out of Wealth, Fame, and Originality, with a maximum of one point in any one area. Furthermore, the last 0.5 points beyond 2 points will almost always involve some type of cheating. Although for many people factors like pleasure and sociability will actually prevent moving beyond 1.5 or two points. Alternately, a maximum total of 3.5 is given, with two extra statistics: Beauty, and Fun, added, which also each have a maximum of 1 point. *"Beauty is great for making up for other things. But other things are not always what people want. On the other hand, great beauties are often known to be monsters in one way or another. Sometimes they lack feeling. Sometimes they are homicidal. Sometimes they are all about being cold and unfeeling, like mathematicians who lost their humanity. Sometimes they are idealistic to the point of scaring people. Sometimes they may be perverts or have no effective communication strategy. Sometimes they may kill too many babies or masturbate too much. Sometimes they may be very poor, or have physical trauma to their body. Sometimes they may have bad luck or be surrounded by evil people. Sometimes (maybe often) they are just stupid beyond words or incapable of speaking the right level of language. All in all, there is not a lot to recommend beauty any more than anything else, all things considered important. The weakness I see in arguing against beauty is that if a great beauty doesn't take drugs and looks abnormal in some way in spite of the beauty, that may give them advantages which drug addicts don't have, assuming the drug addict is not hideous. However, if the beautiful person is also talented, some of the advantages of beauty go away. Although because it may be somewhat relative, the beautiful person may be looked at as achieving a lot in a similar but perhaps different way to the way narcissists sometimes 'collect people'."* —<u>Why is beauty essential?</u>

THE EQUATIONS AND SOLUTIONS TO UNSOLVED PROBLEMS

MISCELLANEOUS EQUATIONS

Nathan Coppedge

,,,

THE EQUATIONS AND SOLUTIONS TO UNSOLVED PROBLEMS

HIGHEST FORMULAS:

Abstract if Results - Min Efficiency ≤ 0

Material if Results - Min Efficiency > 0

Flying if $[((\text{Average D}/2)-1) - 0.5] > 0$

On the Surface if $[((\text{Average D}/2)-1) - 0.5] = 0$

Underground if $[0.5 - ((\text{Average D}/2)-1)] > 0$

"Impossible = New Constant."

"New Universal or Infinity = New Genera."

"New Finite or Possibility = New Species."

"Potential Infinite = New Function."

    ---<u>Official Meta-Concept</u>

Meaningful meaning is 4-d semantics.

If there is no other 'probably' the matter may be resolved eventually.

With sufficient rarity there could be sufficient existence.

Everything might amount to 'people being helped'. Unless they make too fine a point about it.

Just because I like to scare people, doesn't mean I like to scare people about everything.

Nathan Coppedge

UNIQUE UNIVERSE FORMULAS

CONCEPTUAL DIMENSIONS = D

TOTAL MATHEMATICAL CATEGORIES = TC

PHYSICAL DIMENSIONS, (Example of output: 11) = [(Tcategories - Nroot of T) - [(Nroot of T - N) ^N]] = ((TC-(POWER(TC,(1/CD))))-((POWER(TC,(1/CD))-CD)^CD))

NUMBER OF IDEAS (Example of output: 20) = Tcategories - Nroot of T = (TC-(POWER(TC,(1/CD))))

HIGHER LEVEL---> ALTERNATE IDEAS -2 = =(((TC-(POWER(TC,(1/CD))))/4)+1)

ALTERNATE IDEAS -1 = (((TC-(POWER(TC,(1/CD))))/2)+1)

ALTERNATE IDEAS +1 =(((TC-(POWER(TC,(1/CD))))+1)*2)

LOWER LEVEL----> ALTERNATE IDEAS +2 =(((TC-(POWER(TC,(1/CD))))+1)*4)

SPECIES EQUATION: (TC-((POWER(TC,(1/D))-D)^D)+1)

SPECIES EQUATION - 2 = (((TC-((POWER(TC,(1/D))-D)^D)/4)+1)

SPECIES EQUATION - 1 = (((TC-((POWER(TC,(1/D))-D)^D)/2)+1)

SPECIES EQUATION +1 = ((TC-((POWER(TC,(1/D))-D)^D)+1)*2)

SPECIES EQUATION + 2 = ((TC-((POWER(TC,(1/D))-D)^D)+1)*4)

TECH ESSENCES FORMULA = (((POWER(TC,(1/D))-D)^D)+(POWER(TC,(1/D)))+1)

MEANINGLESS CATEGORIES = IFS((TCat<16),"(1) WORMHOLES",(TCat<160),"(2) WORMHOLES, MEANINGLESSNESS",(TCat>=160),"(3) WORMHOLES, MEANINGLESSNESS, AND MORE MEANINGLESSNESS")

THE EQUATIONS AND SOLUTIONS TO UNSOLVED PROBLEMS

UNIVERSAL PREDICTOR CODE =

code part 1

IFS((((TC*(((POWER(TC,1/D)))-1)/2)/D8)+((((POWER(TC,1/D)))-1)/2))<=0,"CELEBRATING CHRISTMAS",(((TC*(((POWER(TC,1/D))-1)/2)/TC)+((((POWER(TC,1/D)))-1)/2))<=1,"PLEASURE",((TC*(((POWER(TC,1/D))-1)/2)/TC)+N(1)+((((POWER(TC,1/D)))-1)/2))<=2,"SEEKING ADVANTAGES",((TC*(((POWER(TC,1/D))-1)/2)/TC)+N(1)+((((POWER(TC,1/D)))-1)/2))<=3,"LANGUAGES",((TC*(((POWER(TC,1/D))-1)/2)/TC)+N(1)+((((POWER(TC,1/D)))-1)/2))<=4,"OPTIMIZATION",((TC*(((POWER(TC,1/D))-1)/2)/TC)+N(1)+((((POWER(TC,1/D)))-1)/2))<=5,"GENIUS",((TC*(((POWER(TC,1/D))-1)/2)/TC)+N(1)+((((POWER(TC,1/D)))-1)/2))<=6,"GREATNESS",((TC*(((POWER(TC,1/D))-1)/2)/TC)+N(1)+((((POWER(TC,1/D)))-1)/2))<=7,"DRAGONS TREASURE",((TC*(((POWER(TC,1/D))-1)/2)/D8)+N(1)+((((POWER(TC,1/D)))-1)/2))<=8,"INFINITY",((TC*(((POWER(TC,1/D))-1)/2)/TC)+N(1)+((((POWER(TC,1/D)))-1)/2))<=9,"PORTALS",((TC*(((POWER(TC,1/D))-1)/2)/TC)+N(1)+((((POWER(TC,1/D)))-1)/2))<=10,"CHAOS",((TC*(((POWER(TC,1/D))-1)/2)/TC)+N(1)+((((POWER(TC,1/D)))-1)/2))<=11,"EMOTIONS",((TC*(((POWER(TC,1/D))-1)/2)/TC)+N(1)+((((POWER(TC,1/D)))-1)/2))<=12,"GOD",((TC*(((POWER(TC,1/D))-1)/2)/TC)+N(1)+((((POWER(TC,1/D)))-1)/2))<=13,"CELEBRATING CHRISTMAS")

code part 2

IFS((((TC/(((POWER(TC,1/D)))-1))/2)+((((POWER(TC,1/D)))-1)/2))<=0,"CELEBRATING CHRISTMAS",(((TC/(((POWER(TC,1/D)))-1))/2)+1+((((POWER(TC,1/D)))-1)/2))<=1,"PLEASURE",(((TC/(((POWER(TC,1/D)))-1))/2)+N(1)+((((POWER(TC,1/D)))-1)/2))<=(N1+2),"SEEKING ADVANTAGES",(((TC/(((POWER(TC,1/D)))-1))/2)+N(1)+((((POWER(TC,1/D)))-1)/2))<=(N1+3),"LANGUAGES",(((TC/(((POWER(TC,1/D)))-1))/2)+N(1)+((((POWER(TC,1/D)))-1)/2))<=(N1+8),"OPTIMIZATION",(((TC/(((POWER(TC,1/D)))-1))/2)+N(1)+((((POWER(TC,1/D)))-1)/2))<=(N1+9),"GENIUS",(((TC/(((POWER(TC,1/D)))-1))/2)+N(1)+((((POWER(TC,1/D)))-1)/2))<=(N1+10),"GREATNESS",(((TC/(((POWER(TC,1/D)))-1))/2)+N(1)+((((POWER(TC,1/D)))-1)/2))<=(N1+11),"DRAGONS TREASURE",(((TC/(((POWER(TC,1/D)))-1))/2)+N(1)+((((POWER(TC,1/D)))-1)/2))<=(N1+12),"INFINITY",(((TC/(((POWER(TC,1/D)))-1))/2)+N(1)+((((POWER(TC,1/D)))-1)/2))<=(N1+13),"PORTALS",(((TC/(((POWER(TC,1/D)))-1))/2)+N

$(1)+(((((POWER(TC,1/D)))-1)/2)) <= (N1+14),\text{"CHAOS"},(((TC/(((POWER(TC,1/D)))-1))/2)+N(1)+(((((POWER(TC,1/D)))-1)/2)) <= (N1+15),\text{"EMOTIONS"},(((TC/(((POWER(TC,1/D)))-1))/2)+N(1)+(((((POWER(TC,1/D)))-1)/2)) <= (N1+16),\text{"GOD"})$

<u>Unique Universe Formulas</u> (...)

THEORY OF EVERYTHING:

Set 0 > Efficiency* + Difference, where efficiency is < 1 if topic is acted on, and efficiency is > 1 if topic is acting. Set 0 = case being considered.

Ranking similarly well-qualified systems: D ^ results > verbs - 1. (D = number of relevant spatial dimensions).

The number of big ideas equals 20 X Genera X Species X Function (Premier Intellectual Dialectic). Revised: 20 X Qualifier: lowest perfect standard: 32 ideas. Perfect imperfect standard: 64 ideas. Perfect ancient math: 2 Ideas. Perfect imperfect ancient math: 20 ideas. Perfect philosophy of mathematics: 100 ideas. Perfect imperfect philosophy of mathematics: 200 ideas. Imperfect mathematical philosophy: 2000 ideas. Perfect philosophy ideas: 10. Perfect imperfect philosophy ideas: 20. Imperfect philosophy ideas: 128.

Impossible = New Constant.

New Universal or Infinity = New Genera.

New Finite or Possibility = New Species.

Potential Infinite = New Function.

A Kind of Universal Universal Constant = Min 0.01706667 to Max 0.0256 exact.

TOE: Results >= Efficiency + Difference

Efficiency > Results - Difference

Difference > Results - Efficiency

Certainty = 0 [Proof: Certainty = TOE - Antitheory = Eff + Diff -

THE EQUATIONS AND SOLUTIONS TO UNSOLVED PROBLEMS

Diff - Eff = 0].

Uncertainty = Antitheory [Proof: Antitheory + Certainty = Antitheory].

General OU = {Efficiency / [(D - 1) (Efficiency)] } + Difference

Anti-Energy = 1 - (D + Difference)

Anti-Theory <= Difference - Efficiency

Anti-Efficiency = Difference - Results

Anti-Difference = Efficiency - Results

Ideal Elements = D + 2

Ideal Principle = Negative [(D - 1) ^ 2]

Un-Ideal Elements = 2 - D

Un-Ideal Principle = [Sq rt 2 (1 - D)]

Basic Meaning = 5/32 proportion

Incomplete Meaning = 160 (constant)

Math Form = 0.10 X Absoluteness

Above Math = Qualities X 10

Languages up to 3-d = 1.585 X 1.09 ^ (D - 2 minimum zero)

Anti-Languages = [1.09 rt of (2 minus D)] / 1.585

Elements - 5 = Immortal Languages

Elements - 4 = Languages.

Elements - 3 = Knowledge.

Elements - 2 = Ordinary Objects

Elements - 1 = Perpetual Motion (ordinary difference + 1)

Elements = Flying Machines Level 1 (ordinary difference + 2)

Elements + 1 = Flying Machines Level 2 (Flying machines used as an element of a grounded perpetuum mobile).

Elements + 2 = Flying Machines Level 3 (Flying machines flying. A large number of flying machines used as a substrate for other flying perpetuum mobile).

Possible Measure of X / Missing-X in Y / Not Y

2 / Avg Speed = Observed (Theoretical)

[Sq rt of 0.5 (Time)] / Avg Speed = Detected

The first observer aims to refute. The second observer acts passively. The first particle responds quickly. The last particle is a slave.

Dimensions - Antiforces = Forces.

Dimensions - Forces = Antiforces.

Dimensions = Forces + Antiforces.

$(D^2 +1)^2$ = Grand Theory number [Former theories: (D (Elements(D+1))+1) OR ((Elements * Forces)+1)]

Subjectively God exists? Default for me is 0 [0 = false, 1 = true]

Subjectively psychopathy? Default for me is 1 [0 = false, 1 = true]

Opportunity for PMMs? [subjectively god exists + subjective psychopathy <=1, then "TRUE"]

Possibility of Tiny PMMs [If subjective psychopathy > subjectively God exists, then "MAYBE"]

Disintegral = - (Difference - Efficiency).

Special Value Theory = [1 (Efficiency) + 0.5 (Difference)] - D

Anti Disintegral = Efficiency - Difference = Antitheory

THE EQUATIONS AND SOLUTIONS TO UNSOLVED PROBLEMS

Antivalue Theory = (Dimensions+(Difference / 0.5 not a mistake) - Efficiency)

Necessary dimensions: This is constant (5) for our universe

Ideal contradictions: (10) Has been found = Necessary Dimensions X 2

Simple disintegral = 1 / Antivalue

Nodes (simpleforms) = Minus Antivalue

Verbs = Diff + 5

OU Formula for TOEs = [(Dimensions ^ Results) - Verbs - 1]

...

Exponential Efficiency = Efficiency - Difference

Fine Things Number = IFS(Diff<0,((Dimensions+2)-((2*Diff)+1)),Diff=0,(Dimensions+2),Diff>0,((Dimensions+2)+((2*Diff)+1)))

Improved formula for the number of archetypal ideas related to something: Disintegral * Fine Things Number: Gives 20 for categorical deduction:

- - (Diff - Eff) * {IFS(Diff<0,((Dimensions+2)-((2*Diff)+1)),Diff=0,(Dimensions+2),Diff>0,((Dimensions+2)+((2*Diff)+1))}

Helium - boron = immortal languages

Lithium - boron = immortal meta-languages.

Byrillium - boron = immortal knowledge languages.

Boron annihilation = ordinary immortal languages.

Carbon - boron = perpetual motion immortal languages.

Nitrogen - boron = flying machine immortal languages.

Element beyond nitrogen - boron = immortal super advanced

flying machines languages.

Elements - 4 = Languages.

Hydrogen - byrillium = super-advanced immortal languages.

Helium - byrillium = immortal meta-languages.

Lithium - byrillium = meta-languages.

Byrillium annihilation = knowledge of languages.

Boron - byrillium = ordinary languages.

Carbon - byrillium = perpetual motion flying machine languages.

Element beyond carbon - byrillium = super-advanced flying machine languages.

Elements - 3 = Knowledge.

Hydrogen - lithium = Linguistic knowledge.

Helium - lithium = Philosophy.

Lithium annihilation = Ordinary knowledge.

Byrillium - lithium = Perpetual motion knowledge.

Boron - lithium = Knowledge of flying machines.

Element beyond boron - lithium = Knowledge of super-advanced flying machines.

Elements - 2 = Ordinary Objects

Hydrogen - helium = ordinary knowledge.

Helium annihilation = completely ordinary.

Lithium - helium = perpetual motion machines of the 2nd kind.

Byrillium - helium = ordinary flying machines.

THE EQUATIONS AND SOLUTIONS TO UNSOLVED PROBLEMS

Element beyond byrillium - helium = ordinarily advanced flying machines.

Elements - 1 = Perpetual Motion (ordinary difference + 1)

Hydrogen annihilation = perpetual motion of the 2nd kind.

Helium - hydrogen = super-perpetual.

Lithium - hydrogen = exemplary perpetual motion flying machines.

Elements after lithium - hydrogen = exemplary advanced perpetual motion.

Elements = Flying Machines Level 1 (ordinary difference + 2)

Elements as they are = flying machines.

Elements + 1 = Flying Machines Level 2 (Flying machines used as an element of a grounded perpetuum mobile).

Antibyrillium + hydrogen = immortal language of advanced flying machines.

Antilithium + hydrogen = language of advanced flying machines.

Antihelium + hydrogen = knowledge of advanced flying machines.

Antihydrogen + hydrogen = Advanced flying machines.

2-hydrogen = advanced flying machines

Elements + 2 = Flying Machines Level 3 (Flying machines flying. A large number of flying machines used as a substrate for other flying perpetuum mobile).

Anticarbon + lithium = immortal language of ultra-advanced flying machines.

Antiboron + lithium = language of ultra-advanced flying machines.

Antibyrillium + lithium = knowledge of ultra-advanced flying machines.

Lithium annihilation = ultra-advanced flying machines.

Antihelium + lithium = exemplary perpetual motion.

Antihydrogen + lithium = advanced flying machines.

Hydrogen + lithium = super advanced flying machines.

2-lithium = ultra-ultra advanced flying machines.

Possible Measure of X / Missing-X in Y / Not Y

ANTIFORCES:

Forces + Antiforces = Dimensions

D - N antiforces = Dimensions of force

D - Dimensions of force = N antiforces

1X Antiforce Min: (D + 2) / 2 - 0.5

1X Antiforce Max: (D + 2) / 2

1X Antiforce norm: (Antiforce Max + Antiforce Min) / 2

BLACK SWANS:

[1] 2.2B.— High fuel efficiency. [RETROFITTERS] [2] 3.1B.— Flying cars, visual interface. [COSMOPOLITICIANS] [3] 2.3B.— Miracle drugs. [SYNAESTHETES] [4] 3.2B.— High-speed communications. [METER HEADS] [5] 2.1B. — Industrial magic. [THE SHOW] and 1.3B.1B.— Life factors, magic factors, experience machine. [DIABOLICS] [6] 1.3B. — Perpetual motion, omniscience, etc. [THE SWEET SPOT] and 3.3B.— Theory of Everything, Anti-Theory, Universal Energy, etc. [UNICOSMICS] [7] 1.3B.2B.— Divine Computing

THE EQUATIONS AND SOLUTIONS TO UNSOLVED PROBLEMS

(Technological Divination). [NETES, PROS] [8] 1.3B.3B.— Making Source. [RESONATORS]

Black swans are not always more radical than radical, unless you're not a radical.

INFERENCE:

If you rank all knowledge using the formula, (D ^ Results) - (Verbs - 1) >= 1 Where verbs must be >= results, you will find many knowledge systems either with low dimensions or high verbs fail to be sufficiently absolute.

PARALLEL REALITIES (?):

Time-travelers know it's one very inspired world (debateable), with one possibility for each human body in each moment, equal to its configuration.

PRIMARY:

Complexity: efficient production.

Efficiency: that which entails paradigmatic results.

Perfection: Insufficient inefficiency with matter.

Incompleteness = high Set 0 or high difference or high positive *efficiency*.

Difference: another topic, or a topic minus its efficiency.

Impossibility: difference with an undefinable value.

SURVIVAL:

If we can fulfill our desires without sacrifice, and make no additional sacrifice, and be authentic enough, and a sufficient number are ambitious, then that is not so bad.

AVOIDING WORLD WARS: Basically if you don't have problems

with allies, and your government is not mishandling things, and there is no threat of invasion, and there is no destruction of major cities, then there will be no world war.

The more people the more problems we can solve. Although if people are better at solving problems a smaller number is adequate, generally a larger number can solve more difficult problems relative to their efficiency.

LIFE:

Pleasure is possible and worth fighting for.

We can get good values with hard work.

We should try to achieve the good life.

In principle when there is a problem we can see life in terms of archetypes.

If we make our archetypes better, life will probably improve eventually.

THE EQUATIONS AND SOLUTIONS TO UNSOLVED PROBLEMS

TRUTH:

Understanding philosophy: I find much is simplified by focusing on *meaning* and *universals* together.

If you add a perfect premise to a perfect argument, the argument doesn't get worse on average.

The luxury platform, and any metaphor for the luxury platform is the difference between thinking and truth.

The cause of the meaningless when it happens: The choice between the absence of a bad meaning, and the absence of a good meaning.

Calculus of Subjectivity: There is an element of truth in everything unless there is an element of subjectivity in something.

Ideal of a Thing: Worthfulness.

PLEASURE:

Principle of Paradise: Success seems to depend on undetermined truth plus philosophy.

Dimensional Pleasure Principle: To be miserable past the '20.5th' dimension you need to have a really good theoretical justification of why/how to be miserable. Thus, most beings in the 21st dimension originate in lower dimensions or have a good or bad theory of how to be 21st dimensional or even higher dimensional. (Whence comes the expression that pleasure originates with theory).

PLATONISM:

Basic Platonic Heuristic: The type of metaphor qualifies the type of justice. Soulful: material. Psychology: reductive. Sinful: truthful. Divine: wisdom, recollection. Etc.

Everything involving Plato involves philosophy.

The hard lesson of eternity is to be humble.

Who one knows is likely a god to one, even if not a confessed god.

One is always worse in some way than others appear in that way.

Theory = Random, means that theories are by nature anti-quantum, anti-physics. Reality is not theory.

Generosity : *not suicide per se*, That one should admit one's failure and remain humble, and all progress follows from there.

That Love is physical, but physics is very great, and in fact beyond metaphysics. That all that matters is love.

That values are tautologies, that is, standards of reality, which requires love and physics and humility and no theories.

That as such, one lives in a kind of very large cave, and if one wills to leave it, given enough time, reality will change, and theory will change, and life will be different than preview.

The Socratic model begins with negation, that is, non-anti-physics and similar principles. This is the very beginning of real metaphysics.

So you see that history is really divine eschatology.

Exceptions are extracted from actualities.

THE EQUATIONS AND SOLUTIONS TO UNSOLVED PROBLEMS

There is always a theory like souls. Souls have good theories. God, immortality, love, intelligence, wisdom, minds, brains, souls, travel, perpetual motion, art, science, the wheel, humanitas.

By the time you time-travel, it's as if it is a form of political training.

When the History has an Epiphany of Parts, Soul.

A God is older than a lie. A God is a history of a soul.

The general categories of categories that may be formed is roughly: 1. Universal category. 2. Extended category. 3. Identical category. 4. Many identical categories. 5. Different or opposite category. 6. Variety of categories. 7. Ambiguous category. 8. Logical category. Now, if we permute these with all manners of knowledge, research, imagination, medium we get a sense of the full variety of categories.

SOLUTION TO PARADOXES AND PROBLEMS:

The polar opposite of every word in the best definition of the original problem in the same order as the corresponding original words.

The words 'problem' and 'solution' must be included if it is not a true paradox, or it is the same as the 4-category deduction when both statements are taken together.

All true general solutions solve all specific problems, and all true specific solutions solve all general problems, therefore problems are exponentially solved or there is a problem with truth.

ABSOLUTES:

Complexity is the fully realized imagination.

Absolutism has a quality of absoluteness.

Relative relativism implies absoluteness.

Knowledge can be measured in degrees.

The absolute life is absolute.

The objective measure is the absolute measure qualified by semantics.

ABSOLUTE KNOWLEDGE:

Note: objective knowledge uses polar opposites in diagonally-opposite positions. They are read recursively counterclockwise by convention of the Cartesian Coordinates, but the recursive reversal is included, so clockwise and counterclockwise systems are interchangeable and only affect which separate deduction is mentioned first.

4-category deduction:

AB:CD and AD:CB

Note: Nouns (persons, places, things, and ideas) and qualities (verbs, adjectives, and substances or elements) are taken as equivalent. In my vocabulary the term 'property' can refer to logical or other neutral terms, or it can be specified as 'noun property' or 'verb property'. The exceptions to this may be considered physico-dynamic (expressed in logic) or conceptual (attached to the matter: subject to the results of a system of sufficient qualification). Note that all deductions within a set are equally valid given coherence (hence the AND symbol), and usually may apply simultaneously. They always apply simultaneously given coherence.

THE EQUATIONS AND SOLUTIONS TO UNSOLVED PROBLEMS

16-category deduction:

[A] ABFE-CDHG-KOPL-IMNJ
[B] ABFE-CGHD-KOPL-IJNM
[C] AEFB-CDHG-KLPO-IMNJ
[D] AEFB-CGHD-KLPO-IJNM

64-category deduction:

(1) 1A2A 3B4B
(2) 1A2B 3B4A
(3) 1B2A 3A4B
(4) 1B2B 3A4A
(5) 3A4A 1B2B
(6) 3A4B 1B2A
(7) 3B4A 1A2B
(8) 3B4B 1A2A

1A:

1,2,10,9 | 3,4,12,11 | 28,27,19,20 | 26,25,17,18

1B:

1,2,10,9 | 3,11,12,4 | 28,27,19,20 | 26,18,17,25

2A:

5,6,14,13 | 7,8,16,15 | 32,31,23,24 | 30,29,21,22

2B:

5,6,14,13 | 7,15,16,8 | 32,31,23,24 | 30,22,21,29

3A:

64,63,55,56 | 62,61,53,54 | 37,38,46,45 | 39,40,48,47

3B:

64,63,55,56 | 62,54,53,61 | 37,38,46,45 | 39,47,48,40

4A:

60,59,51,52 | 58,57,49,50 | 33,34,42,41 | 35,36,44,43

4B:

60,59,51,52 | 58,50,49,57 | 33,34,42,41 | 35,43,44,36

Additional detail: Levels of neutral Boolean operators for categorical deductions: "is", "as", "just as" "when, "so", "and as such", "just that", "such that such that", "grows", "parse me such that", "incomprehensible unless", "says that", "possible when", "outwardly"…

ADDITIONAL DIMENSIONS:

Dimensional Existence: Soul distributed over a tetrahedron in the case of hypercubes.

Meaning is the measure of all things.

Piece of Nirvana from God = superpower.

Doubt under the command of God = 4-d.

Key to Hypercube: Categories cannot be rotated, or it is ambiguous how they are located: IFNOT, THEN 4-d has 2-d sideways depth.

If you can add significant usefulness you may as well add a dimension eventually.

THE EQUATIONS AND SOLUTIONS TO UNSOLVED PROBLEMS

EXCEPTIONS:

In any moment, the universe has already sustained maximum damage.

The major alternative to discrete values is universal dynamism.

Bullcrumbs are an example of anything.

TECH INNOVATION:

Primary Innovation Formula: A (C ← → D) not = B, for any (A, B, C, D) in terms of A and B, in which A is polar opposite of C, and B is polar opposite of D, and polar opposites are diagonally opposed. The only exception to this appears to be evolution already assumed under a specific product name.

Alternative to Digital Displays: Perfect image = perfect logic.

If philosophy is measured by specialization, wealthier economies tend to be more philosophical.

PROOF THEORY:

Some U := U

U : = Some

(U : = Some) := Some U := U

Where Law is Some, Law = U

Where the universal standard is X logic, the universal proof derives from it (Alternative Systems).

MISCELLANEOUS LOGIC:

Macro System

= profound dualism plus difficult demonstration

String theoretic deduction: X → String theory → Space Time

Female calculus (abbreviated): Neo-Platonism, I think I found Nihilism → I'm the shit*, I know everything, how's my posterior? (*Meaning she believes in God).

Rule of Invention: Invention is not psychological unless it is a psychological invention.

Formal Planar Addition (Concept): Terms are added in terns of equality. Two things are equal if they are equally coherent and exist in the same hierarchy, or they may add hierarchies if there is application. Alternate approaches may be possible, for example, terms may be added if they are equally incoherent and also both contingent to the same hierarchy, and addition may not be possible if terms are duplicates, except incoherently.

Iterative Wave Theory: The iteration of all capacities, present at all levels.

Its like knowledge to like something besides knowledge.

In quadra, Paradigm A B <-> D Advantage C.

Solution to black holes: Is it coherent?

Disproof has one degree of contrast, and contrast has one degree of reflection, therefore disproof has one degree of truth.

Algemeine Equivalence: Protophysics on the model of the Algemeine.

THE EQUATIONS AND SOLUTIONS TO UNSOLVED PROBLEMS

Why squares? Because circles would have to be part of a typology. Why squares? Because squares reduce to axes. Why squares? Because they are axiologically parsimonious compared to triangles, and hexagonal are just 3-d squares.

POSSIBILISM:

Free Will = scale and realized quantity of preferences.

Most things that exist are possibilities that are not actualized, unless life is both perfect and complex.

Actualize the universe? Probably we'll either spend too much energy, die out, or declare the result meaningless, or we're already halfway there. Aim a little past halfway and you see possibility.

Infinite knowledge is possible because the standard of infinity can change, and prior work may be more arbitrary: when arbitrariness decreases, and standards of infinity change, infinite knowledge is possible.

The determination of whether the universe is objectively good or bad may be arbitrary, because good and bad are both objective concerning themselves.

Primary formula for free will: Or And And.

ETHICS AND DECISION-MAKING:

Principle of the Good: It might be judged to be very good that things are not absolutely evil. For how would things be if this were not so?

X + Metaphorical Epoxy = Ethics

Rule of Essence: There is little that does not conditionally exist. Most of evil is a condition. Ethics should aim to affect conditions.

Trivial really bad responsibilities: If something is bad enough and trivial enough and within human hands, it will not occur.

Ambiguous Choices: Whatever predicts general improvement should be preferred, even if the knowledge is not easily secured.

Ethical responsibility: Is fated, creative, or corrupted.

Real responsibility is a responsibility to be perfect, which is only possible if one is already perfect in some way. However, it may be imperfect to believe this.

The winning theory is ethics, according to which the universal is good.

Where something is meaningless it is because there is a choice between meaning and the good, therefore the meaningless is unethical.

If it is logical to act ethically, then mathematicians act ethically insofar as they are logical, or there is some way math is not associated with logic. Escape Strategy 1: It is not logical to act ethically. Escape Strategy 2: Math is not ethical logic.

THE EQUATIONS AND SOLUTIONS TO UNSOLVED PROBLEMS

SELF-DEVELOPMENT:

Formula for an interesting life: Philosophy and its exponents.

AESTHETICS:

Meaningful concept and meaningful representation are never fully unified, as concept is not representation.

Art without technology by some definition is not designed. The type of 'technology' shows the type of design, yet design by itself is not concept.

Beauty: Meaning involved with understanding.

Advanced Design = specialized & extended.

LANGUAGES:

Pidjin Equation: People can be different.

Spanish Equation: You will need only several expressions and suffixes to sound slightly intellectual in Spanish.

Chinese Equation: Note: this is a strange, alternate system you are not likely to learn in school. 1. Practice art. 2. Practice science. 3. Learn Chinese. 4. Process of elimination. 4. Master all arts. *Advanced Chinese:* Chinese is a vertical language. It has many nuances which are not immediately necessary. This was the original idea of how to learn language rapidly…Everyone who learns Chinese is an honest scholar. *Rumors:* Some say Chinese is a simple language, but there are many variations. Some say, the less you practice the better you are at Chinese. Some decide this is false. Some claim that ancient Chinese is the only Chinese. But, there are some modifications. They say Chinese is the language of China. A lucky language.

Equation of Basic German: Its like runes, and you write small things around it.

Equation of Arabic: The order of characters explains how the characters are to be understood, not read. For spoken language, concepts, rather than words, are memorized.

Prevalent Theories Theory: That the legitimate view of a theory is the one communicated by the language.

TIME-TRAVEL:

Backward Travel: Split Existence.

Backward Travel: Uniqueness + Guarantee.

Forward Travel: Share your luck.

Forward Time: Singularity.

THE EQUATIONS AND SOLUTIONS TO UNSOLVED PROBLEMS

GENERAL PERPETUAL MOTION:

Volitional Equations:

Volitional Energy = Mobile U / Necessary Dual-Axial U

Volitional Equilibrium = Modular U / (stems per cycle / subcycles per cycle)

Volitional Efficiency = Volitional Energy / Volitional Equilibrium

The vertical begins at 45 degrees.

Official General Theory:

I estimate the barriers are 1, 1.5, 2, >2.

At unity, energy can continue with zero resistance.

At 1.5 advantages may be had. For example, 1.5X leverage, 1.5X mass.

At 2X you can lift a weight with an equal weight, probably using >1.5X leverage vs. >1.5X mass to create 2X efficiency.

At >2X you get very exceptional efficiency, but this only might occur with exceptional design.

As a result of this, I conclude that a real perpetual motion device will have about 200% efficiency arising from 1.5 to 2.5 ratio of weight versus leverage, but all this means is it can lift its own weight.

>= 50% Vertical = dual-directional (bad).

Miscellaneous:

Magnets are never necessary with real perpetual motion.

It does not guarantee high energy return.

Rather, energy return will be directly scaleable to mass.

If there is matter without energy, then there is matter from energy. Bad ideas sometimes work or physics does not always work → Perpetual motion might work.

A speed of one unit per unit might be conservative if it defines conservation.

Expressed in ratios, from a minimum of compensating for max leverage * >0.5 effective mass + effective long-end lever mass unweighted to a maximum of max leverage + effective long-end lever mass unweighted = Approximate exact range of working counterweight mass in all devices. For example, for a 2:1 lever the counterweight will be about 1X mass + lift + weight of lever, whereas in a 3:1 ratio the counterweight will be about 1.5X mass + lift + weight of lever. These values will yield ranges of >1.5 rising ball : 2 counterweight : >2 falling ball and >2 rising ball : 3 counterweight : >3 falling ball respectively. Thus increasing leverage by 50% results in almost 100% increase in the allowable mass ratio for the lever when it is unweighted on either end, creating an efficiency.

Rule Responding to Heat Machines: Perpetual motion of the first kind requires a low-heat system to cut heat losses, but there needs to be at least a small amount of excess heat if you're creating energy.

Assuming ball = 1 with variable application, and long end = additional 1 constant application, Max counterweight mass = less than the minimum amount of leverage + 1, Min counterweight mass = greater than (highest amount of leverage +1) / 2, Unified Counterweight Mass Formula = Min + 1 > (Max + 1) / 2

SPECIFIC PERPETUAL MOTION:

Leverage Class, Max counterweight mass = less than the minimum amount of leverage + 1

THE EQUATIONS AND SOLUTIONS TO UNSOLVED PROBLEMS

Leverage Class, Min counterweight mass = greater than (highest amount of leverage / 2) + 1

Leverage Class, Unified Counterweight Mass Formula = Min Lvg + 1 > (Max Lvg / 2) + 1

Heavy Ball Class, Min ball bearing mass = greater than (long-end leverage + 1) / 2

Heavy Ball Class, Max ball bearing mass = less than long end leverage +1.

Heavy Ball Class, Unified Heavier Ball Bearing Mass Formula = Long end leverage +1 > ((long end leverage +1)/2)

Friction does not eliminate motion where motion is permitted.

Reactions are possible in a circle, as shown by dominoes. Wheels can turn.

Dominoes can chain-react using higher and higher altitudes. Energy can be created.

Dominoes can accelerate, so friction does not stop everything.

In principle, equilibrium is enough to overcome proportionality problems.

Imbalance can overcome friction.

Equilibrium and imbalance can exist simultaneously through mass-leverage ratios.

With unbalance and a principle of momentum, there is no need to lose altitude over time.

All else considered, natural momentum with no net loss of altitude = perpetual motion.

With natural momentum and upward and downward motion, potential for return.

1st Fully Provable Usable Angle: When theta end of lever >= horizontal inverse of theta track slope,

The Escher Machine: Theta (Delta H) > Theta (V) . In other words, the wedge in this machine makes gains on the difference between the horizontal and vertical angle using momentum from the backboard.

Nathans Divine Perpetual Motion: 3-d Semaphore + 4-d Wheel.

Natural Torque Device: Natural torque, without net altitude loss, expressed in mass, with recoverability.

First Fully Provable problems: Basically, very close to the exact ratios must be kept for it to work, but the masses are flexible when we keep the ratios. Therefore the counterweight is not really flexible by itself, and therefore the leverage ratio must be closely maintained. As a result the mass RATIO is not flexible, and as a result the lever structure unweighted must be kept ulta ultra lightweight. These conditions essentially must be met.

First Fully-Provable Master Equation: Given correct ratios, if the balancing weight is flexible, it's in the bag.

First Fully Provable Maximize for Lightweight: long-end lever mass significantly less than 1/2 ball mass.

First Fully Provable Long-End Effective Mass: Total lever mass divided by long-end leverage ratio.

Real Over-Unity Experiment 2: When altitude gain is maximized relative to the slope, the inward spiral can be extreme and still maintain a steep downwards drop on return.

Swivel Device: If it moves l/r automatically in the same position at the same altitude, that suggests perpetual motion, with other things considered.

THE EQUATIONS AND SOLUTIONS TO UNSOLVED PROBLEMS

MATH:

Limit of the origin: helps with calculus.

Polycalculus: Physics → Number theory → Qualified or unqualified.

Disintegral: Given: The Indefinite Condition, and Arbitrary Structural, Then $\lim(\phi)$ = infinite chain

Universal simulacrum: Correspondence ^ stars

Logistics: Everything in every place, given (sufficient scale, finite space, finite time).

Inflections on greatness is a beautiful combinatrix.

Ordered systems are often better at handling chaos.

Coherently numbers = 1 because that expresses the sum.

There are no coherent varied number strings, because at least half of all expressions are not numbers or values. Coherent strings are not number strings, but rather more like loops, analogies, or Kripke operations.

Probability: Useless 14

Calculus + L/47 : Age at which one can have sex, where L is the standard unit of time corresponding to one's absolute level. For example, if you live a century, and you know calculus, you have sex around age 50. But if you are God it is measured in decades. 10/47 equals about 1/4 of God's years before he makes love. A 13-year-old will eventually have sex at about ($\pi * 3$ / absolute level) years * level of superficiality.

To know one doesn't move, and see things move without moving and time travel and be immortal: 4-dimensional consciousness.

The limit of eventually is a constant.

Lines have perspective unless they are contingent.

PHYSICS:

Make space-time improbable, and add energy = exotic matter

If T is strange approx = time travel with Bohmian Mechanics

"Expensive system" → "everything exotic" = 2-d time

Absolutes can be inverted, but they are non-relative

Cold stars can exist in hot space. But only as a relation or relative dis-relation. Of matter and antimatter = Antimatter energy (this particular from Aug 1st, 2016)

Black holes have memory is the reason that they're empty.

The universe is created any time a black hole isn't plugged.

4-d: Not purposive, permanent.

What is complexity? A self-swallowing solution with a self-swallowing problem. In other words, an open matter with a lemma and a method.

Reassurance: Too much.

What is a little bit different is subtle anti-matter.

> Electricity is 2-d Maxwell.

> Magnetism is 3-d Maxwell.

> Unification of electricity and magnetism is 4-d Maxwell.

THE EQUATIONS AND SOLUTIONS TO UNSOLVED PROBLEMS

Magnetism is 1-d?

Magnetism is time?

Cross-flux spacetime is entropy?

Involution is the reaction between time-magnetism and 2-electricity?

Some credit to K Rubenstein (?) for suggesting electro-magnetism was already unified.

Quantum Physics is Anti-Eternal-Return: The opposite of unified randomness will be a determined singularity. Eternal return requires time-travel but not everyone is a time-traveler, nor will life be determined the same way after it has already been determined, as the quantum state will be different but not opposite.

PROTOPHYSICS:

Sub-information is like a colloquiam.

Algemeine: It means the general theory of algebra, particularly the relation between whole orders of sets and the specific operators involved in defining the algebra.

The standard tag is a modified design.

An image is an off-the-shelf example.

The usual case would be to modify a string.

The purpose has at least one pole.

Complexity is often distrubuted, with an ambiguous meaning.

Whatever is more perfect than complex is plagued with answers.

We have learned the hard-won information that much of meaning consists of notes, and how to use them.

Intelligence is very often a dedicated association, or more rarely a rambling quality embued with luck.

Very often higher intelligence is an idea of how to be an enchanter, or something similarly inspired.

Gods, if there are any, are masters of immortal nature, and problems more difficult than man's.

OCCULT:

Exaggerated high energies = magic

Potentials of low energy = meaning

Fate is in proportional relation to the relative reality of the stars.

If there is such a thing as psychic possession, it is highly inefficient, because if someone was motivated to be you, they would be you.

Exceptional power: rarer than scarier.

Because what is truly improbable is less real, magic exists to the extent of probability. Magic exists to the extent of probability, therefore it exists. Everything exists unless there is a contradiction. Contradiction doesn't exist ethically, metaphysically. Ethically-metaphysically is an exceptional exception. Magic does not by definition equal contradiction, therefore magic exists.

The universal that is beyond knowledge IS knowledge.

THE EQUATIONS AND SOLUTIONS TO UNSOLVED PROBLEMS

HISTORY:

Ancient Greece is as random as 1511.

If we're half progressed we have a right to see halfway into the past, and see the earliest part as meaningless. If we are 3/4 progressed we have a right to see 1/4 of our history, and see the furthest part of the 1/4 as meaningless. The more progressed we are the less of our history we have the right to see, for if we saw it we would also inevitably see the irrational efficiency of being more fundamental.

New radical ideas usually occur at the same time conservatism is the strongest, within about 5 - 500 years of each millennium.

ANTHROPOLOGY:

RISD is not a fossil.

METAPHYSICS:

The Form of Universal Dimensions: Given certain potentials, ideas always exist. However, those potentials are not as reliable as some might assume. So the exceptions of all dimensions exist simultaneously in various proportions.

Nature vs Nurture is like Learning vs Memory: Being reborn in a genius body is about as probable as intelligent strategizing.

3-d = incoherent space.

4-d = incoherent substance.

3-d space + 4-d substance = incoherent universe.

Psychology of Transmigration: Generalists who are admired are not depressed.

Solution to Mary the Superscientist Locked in a Black-and-White Room: If colors exist, they will eventually be observable using concepts in infinite time. Every concept has stress in infinite time.

Symptomatic Reality: Degrees of manifestation are directly symptomatic of the infinite.

Functionality is a corollary of standards.

If the universe is finite this indicates a finite number of categories, which suggests design. If the universe is infinite, this suggests infinite categories, which suggests eternal nature.

ARCHETYPES:

One archetype precedes another.

Two clever things including one simple clever thing before we get the second simple clever thing.

Equivalent of one nano thing and extrapolation before one mini thing.

Equivalent of two mini things or thoughts of combinations before one micro thing.

Equivalent of three micro things or an idea before one meso thing.

Equivalent of four meso things or psychic before one complete set.

Equivalent of four foundational works or one epiphany before a macro thing.

Correct principles, tradition, and effort before one infinite thing.

Two infinite things to create five more infinite things.

Originality and genius for archetypes.

THE EQUATIONS AND SOLUTIONS TO UNSOLVED PROBLEMS

Perfection in some sense for immortality in some sense.

At least immortality for godhood.

Qualified magic for wisdom.

Ongoing fascination for divine intelligence.

Knowledge of distance for power.

Problematics is the ability to solve problems.

Identity is the suitedness to a situation, requiring situations and suitedness.

Inexorable judgment is the ability to tax meaninglessness.

Divine solutions are alethic formulas.

Divine art is an adequate response to ugliness.

What is true has a false and uncertain double.

Integrating the Shadow Archetype:

We reach out to something dead within us…Sometimes it's medicine… Further along, behind empty words… Are the channels of the unconscious… Enormous nakedness…Foolishness…Wordlessness…Beautiful silence…Simple gladness… Contrast… Hope.

ALL OF SCIENCE:

Beau 42.

Rule of Scientific Realism: Prescient or subtract 1300.

Scientific Findings: new + ideas.

Preliminary formula for a Theory of Everything: (5/16) Mind. matter, meaning, woman. (11/16) Describing (infinity, mathematics, calculus, science).

PSYCHOLOGY:

The rule of the soul: The common comes before the good.

Happiness = Notice that you probably don't really feel infinitely blah.

Fairness = You're responsible for what you know.

UX: Preferences → Keywords → Interactions → Inferences → Improved experience → Better paradigm → Refocus → Improved customization → Better products → Better viability → More investment → Improved overhead (not just money, but dedicated resources) → Improved social infrastructure → Better society → Improved preferences.

Neurology of Languages: People learn things that have no negative association, or they forget things they feel happy about.

Mental Illness: Bad options resulting from good determinations.

Possibility arises most coherently from impossibility. --Universal Psychology

THE EQUATIONS AND SOLUTIONS TO UNSOLVED PROBLEMS

COGNITIVE SCIENCE:

Help with [Quality] or something (low-level brain boost).

Put to an incomplete arbitrarily complicated task, however, the brain can be very limited.

Either life is peaches, or you're doomed, or you don't understand psychology.

If you're wrong, then you don't understand psychology.

1000 IQ Formula= Stimulating yourself to be Einstein ^ 14.

Bad theory is worse preferences.

Synapse formation: New synapses are formed from the obvious deformation of the disintegral.

Nature of Neurons: The question of the many kinds of neurons is answered by the singular non-typed genera of the body.

Neural Code: It requires a code.

Neuronal Free Will: The freedom of the sensory system is determined by reactions against systems.

Evolution of the Brain: The question of the brain is answered by 'why not be immature?' The quality of individual brain substances are determined by an indeterminate number of immature groups.

Organization of the Brain: The senses are chemical reactions disintegrating, the rational purpose was imposed through organization, which is a kind of extraneous characteristic, an offshoot of chemical reactions disintegrating.

The brain location of costs-to-benefits is embodied with qualities of double-negative benefits, this has the result of creating brain plasticity.

Dreaming: Curiosity about thinking and curiosity about visual vision.

CONSCIOUSNESS:

Consciousness may be a luxury platform, more rarified than non-existence.

Consciousness may be: (1) A kind of official status that results from redeeming points. (2) A response to extreme stimulii like pleasure, pain, excitement, and feelings of familiarity… an acquaintance with fascination and obsession. (3) A 'brain-trap' of reliance on a particular mode of thought, and particular level of ability to assess and synthesize experience [arising]… from particular ideas. (4) Consciousness may… be whatever physical substance or dynamic that happens to be conscious.

Consciousness: 'The evidence of a superiority to its own vision'.

Consciousness cannot be annihilated: lucky if true and unconditional OR some better theory.

Perception: To reflect some part of a thing's essence.

Thought: Consciousness multiplies by variation.

Solution to the Hard Problem of Consciousness: Unlikely solutions involving consciousness, and likely solutions not involving consciousness. Or, simply, luck (best unlikely theory) involving consciousness, or commonality or shared traits (best likely theory) involving unconsciousness.

THE EQUATIONS AND SOLUTIONS TO UNSOLVED PROBLEMS

IMMORTALITY:

Demigoddery Formula: Historical greatness AOR magic) + immortality.

Human Immortality: Depends on the rules.

Anyone who enjoys themselves wants to live forever, unless they made an evil bargain.

Formula for Human Immortality: Everything is Ambiguous, and then God Dies.

Immortality Def: Everything, OR Sufficient Relation.

POSTHUMANISM:

Without problems, there is less need for average human adaption.

POWER:

Formula for Power: Authenticity → Adaptive prediction → Manifestation → Coherent category → Dedicated category.

THEOLOGY:

God by a lower standard is God by a lower standard.

The first thing is superficial. The second thing is difficult. The third thing is God. If the first two things result in politeness, and there is a God, then the universe is good or we must prefer a different process (probably an authentic one). But if the result is rudeness, there may be no superficial way to please God, and so the universe may be acting rudely. Consequently, if superficiality may be realistic, God may not be realistic, unless there is an easy option for everyone.

Mystery: recorded genius.

If humans aren't divine, the timeless God does not look like a creator, unless humanity has crude origins.

God is work. Summed in Peril.

'ASTROS':

With no violence, there might be knowledge and ignorance which prevents intermixing, suggesting that other aliens will remain distanced until knowledge and ignorance are intermixed.

PHILOSOPHY ABOUT PHILOSOPHY:

In one sense, justice is then anything that is divine. In another sense justice is finding justice, which requires wisdom, and requiring wisdom is the task of philosophers. So, we might predict that philosophers are seeking to become gods, as well as that gods are incapable of sin. This apparently is physically and metaphysically undeniable.

THE EQUATIONS AND SOLUTIONS TO UNSOLVED PROBLEMS

PROVERBIAL EQUATIONS:

Seriousness is central because the absurd is so important (seriousness is important if the absurd is absurd).

Problematically problematic can teach the wise to be wise.

What people think is unquestionable is often a bad visual idea.

A deception that is a metaphor is pointless.

Metaphysical promotions are the objectivity of metaphysical tools.

What has natures comprehends.

Dangerous symbols: alien genders.

EXPERIMENTAL:

Art as a letter from the future.

+/- experience when it swerves. Time is a contingent quantity.

God needs a heart that isn't purple, then he needs takeout.

Form might arise from immortality, or double contingency, or contingent energy, or exponential efficiency.

Some things we don't perceive don't affect this local universe. There are things we're not dealing with that don't matter. There are things we're not dealing with that aren't perceived. There are things we are not dealing with that are radical.

When people are aware we're not supposed to make infinite sacrifices.

Ah, that's the God that I know: the God that I don't know: it's not God's truth goes the theory. So if God is an epiphany it might have legs to walk on. Yet, does epiphany mean God? Well, perhaps not if we must deny God to deny we know Him.

NEXT:

"THE PSEUDO EQUATIONS"

THE EQUATIONS AND SOLUTIONS TO UNSOLVED PROBLEMS

SHORTER DOCUMENT: THE PSEUDO EQUATIONS

(I have thought that some of this is the worst for aliens that are violent:)

If God is a cannibal, something is not human.

Subjective Nirvana: They're just being skeptical & similar.

The secret of Hell is, its empty and it hurts.

Technological children = fatherhood + coincidence and serendipity.

Elitism: It's just elitism isn't it = IDK.

See what is charged for stupidity? Everything, apparently not including stupidity.

Satan is a representation of authentic bias.

Second third infinity of fallacies is what many multiple correspondence relies on, as correspondence is potentially a fallacy which compounds with number of correspondences.

Fortune: Things change when its no big deal.

Depression: Overthinking how to underthink.

Everything about China is prophetic to Lao. If Lao is immortal he will be shackled.

Fashionable: Two minutes of God's time.

Cheaper People: They are more authentic, because they have more freedom or less history.

Fantasy and the coincidence of how you treat everyone: the damage to souls.

What do we trust except authenticity, so you shouldn't assume I'm the one causing the problems.

There's always a complex explanation that works, when lies are involved.

If God has you by the p****, you might think you're God.

It seems if someone wants to send me to hell, life must seem too good to them, they think something is wrong with life and living.

Cubists invented ghosts. The greatest Cubist invented the greatest ghost.

It's madness if it's impractical.

Animistic mind: Irrational personality in a very sane setting.

Let's consider that my complaint could be of weight. True, my (complaint) may have token value. In the information age, token value was played bigtime. Token value may be relevant. Therefore, my complaint may be relevant. Where my complaint may be relevant, you should take my complaint as though it were serious. If that is the case, what does it mean if my complaint is of weight? It could be very weighty indeed!

If we know our aggression we transcend.

If something is really fun to say, you might not care what the truth is.

Ambiguity about irrationality is just sane people messing up.

How to lie like Lucipher: Be honest.

Gods are always-happy people who somehow understand pain.

THE EQUATIONS AND SOLUTIONS TO UNSOLVED PROBLEMS

You see people similar to you no matter what you look like, unless you're an authentic person or a crazy person. Normal people always want to be authentic people. Crazy people think maybe they want to be authentic people. Authentic people are envious of crazy people. Crazy people are envious of authentic people.

Parfit's Formula for Becoming Lucifer: Parfit says either the soul or ideas is real, and we should do something with that (if we can, if we know what to do, unless it has bad consequences).

Thoughts humans smother to death: Sounds violent but it feels soft.

Social karma from talking out loud creates schizophrenic voices.

Unavoidable Survival Principle: One of my guesses is that those who are very rare, but do not find their existence valuable tend to be very tough human beings who have more life left to live. The principle of how consciousness avoids unnecessary experiences.

Nathan Coppedge

,,,

THE EQUATIONS AND SOLUTIONS TO UNSOLVED PROBLEMS

OPTIMISTIC STATE OF KNOWLEDGE

We know the main categories in the 2ToE25 if we study Nathan Coppedge.

We may know less about:

- Extensions of traditional technologies like faster-than-light travel and hallucinogens.
- More efficient chemistry and evocative language has not been developed yet, except in rare cases.
- Pharmaceutical research has not developed to the point of effectively eliminating serious diseases.
- There are numerous other applied fields where traditional research has not been preserved, or has not advanced to the level of perfect understanding.
- There is plenty of work that has also not been done in computational research, such as full investigation of the possibility of omniscience and complete investigation of all technologies.
- Cosmological data.
- 3d ToEs, 4d ToEs, etc: one clue is they may involve $5^{\wedge}D$ categories and additional dimensions of number theory. The additional variable is probably initially D or some type of efficiency or perfection, though it is not necessarily known.
- Alternatives to the ToE including the Efficiency Equation, Dimensional Equation, etc. and related technology. However, Coppedge knows that these equations which are part of Coppedge's Immaculate Equations are often optimized at similar numbers such as 3 for Var1 and 1 for Var2 and a few alternate options. The math has been extensively described an an elaborate manner for all of Coppedge's Immaculate Equations. It is a matter of whether the equations are preserved and re-applied.
- Applied technology. For example, there are some recent technologies that while not a pipe dream are not widely researched such as: perpetual motion machines, Nathan Coppedge's recoilless rifle concept, the gravity hammer, free energy boat and ship concepts, self-powered flying machines, self-powered batteries and lasers, consumer energy generation, waveform effects (luck ray, etc.), and many other areas both explored and un-explored.

Verification of some of the fundamental principles of perpetual motion from science is wanting, though the initial experiments are promising.

THE EQUATIONS AND SOLUTIONS TO UNSOLVED PROBLEMS

PRECEDENTS

Work that may be owed partly originally to Nathan Coppedge:

The invention of immortality. —Nathan Coppedge in past life as Zheng Guo god of immortality—It is inexplicable here that 'god' seems to mean 'inventor' that it may be the original god and that God is rarely referred to as an inventor in the modern day.

Socrates does not appear to have left the precise formula for souls, except through the interpretation of irony.

Leibniz's major approaches to intelligence may evolve into 'Abstract Polyps' for reasons that sound a bit occult:

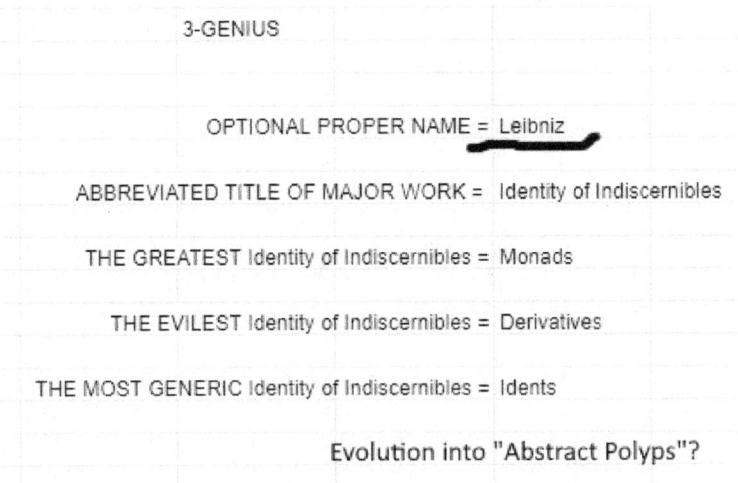

3-GENIUS

OPTIONAL PROPER NAME = Leibniz

ABBREVIATED TITLE OF MAJOR WORK = Identity of Indiscernibles

THE GREATEST Identity of Indiscernibles = Monads

THE EVILEST Identity of Indiscernibles = Derivatives

THE MOST GENERIC Identity of Indiscernibles = Idents

Evolution into "Abstract Polyps"?

Kant does not really attempt to have real metaphysical (generative knowledge), rather he attempts to interpret the world through the lens of ethical principia. Nonetheless, Kant's metaphysical thinking is a major precedent for what I called the <u>Personal Physics</u>

Just a note, Heidegger may have been too emotional to invent irrational metaphysics.

1974. What is it like to be a bat by Thomas Nagel may have led to this work by NCoppedge: <u>How To Think Like...</u> (...)

Quantum Dots (?) (Coppedge, Unity Project from 2004 anticipated use of exponentially-efficient permutation)

Bill Leigh's answer to What are some innovations that will happen in the coming years?
2009, 'hyper-super-meta-ultra' became important for the founding hyper-car in March 2016 as seen on wikipedia now. I think if people think about it the real hyper-car experience did not exist until magazines released an image of this particular model: Koenigsegg Regera - Wikipedia However, while Nathan may have had some influence on the name of these cars and perhaps even some body designs, Nathan cannot place his own influence further back than about 2016 on the term 'hyper-car' in the following writing: If you went back in time 15 years what would you do? Nathan remembers thinking around that time that the terminology 'supercar' was out-dated and that it was unbelievable no one had used the term 'hypercar' even once, or that is what I remember. Or possibly it was available under one spelling but not another. However, 'google trends' says the term existed in March 2004.

2010, Jammers music. Nathan noticed no one would ever think of ninjas hurting their thumbs while listening to pop music.

2011, In a forgotten time-travel event, Nathan may have inspired the "anti-caffeine warrior" with the idea that it was the most masculine thing to 'fight against caffeine' relatively speaking. Nathan called this 'S1911' for "Secret of 1911".

2011, I gave a tip on the 3-body problem that with cubes it is the negative space raised to the vectors of velocity plus impacts. And with spheres in 0-G without impacts it is the relative ratio of mass times the gravitational force in rral time times the inverse of the individual velocity.

Later echoed somewhat here: Physicists get close to taming the chaos of the 'three-body problem'
2012, Equivalence relation that is not a bijection: Exclusivity, Polarity, Analogy: Nathan Coppedge's answer to What is an example of equivalence relation that is not a bijection?

Universal Software Model (?) And preference for simplicity of polar opposites over limits *(Coppedge, Categorical Deduction from Feb 2, 2013 anticipated coherent modeling. Numerous other systems by*

THE EQUATIONS AND SOLUTIONS TO UNSOLVED PROBLEMS

Coppedge have kept the trend going)

Gheorghe Matei's answer to What are some innovations that will happen in the coming years?
Feb 2, 2013: Improved quantum circuits:

"The research team proposes the use of ZX-calculus as a language for this intermediate stage of compilation. ZX-calculus is a 2-D diagrammatic language (using diagrams and imagery instead of words) developed in the late 2000s...... [the] two representations of logical gate circuits can be mapped to one another... The ZX-calculus language can apply a set of transformation rules [*while remaining quantum*] ... and therefore ensuring its correctness... achieving considerable compression rates... Interestingly, it could also be the foundation of future operating system development," said Kae Nemoto... "our method **could save a great deal of effort associated with hardware development. [Is it not likely then that this is a coherent description of all of nature which does not require quantum computers? Who did this research?]**" ... (we made a corollary now we're having a coronary type of stuff) —Physorg: New Approach to Circuit Compression (Nov ? 2020)
Master Angle (Coppedge, July 3, 2014)

Magic Angle in Graphene Verified by Scientists (2019)
Dimensions as a brand (Late 2015)

"lego dimensions"—'lego' company sort of stole some of my fame with that idea. Kids were primed to learn about dimensional philosophy and Programmable Heuristics, but ethicists were against it, and legos stole the term 'dimensional' so it was thought to refer to legos. Their defense is that they never call it 'dimensional' they always call it 'dimensions' but if you ask me this caters to people that don't know what's good for them. —Why do kids like Lego Dimensions so much?
Curved Rail Device (Patented by the Chinese 2015)

Chinese patent CN105587479A - Gravity perpetual motion machine - Google Patents Filed 2015–11–20, I have no legal problem with this although my Curved Rail device was notarized in the U.S. July 7, 2006. See: Chinese Perpetual Motion

Formula tor souls 2016

- Was later anticipated to align with Zero-Energy Information Theory by the theory that perfect knowledge has zero energy.

Disintegral (May 19, 2017):

Advantages on Unified Theory of Quantum Plasma Physics: Title: "Unified Theory of Quantum Plasma Physics" Soul: If you [are coherent enough], [objective enough, cold enough, and abstract enough] you will discover the disintegral [is not mathematical enough] This refers to a specific insight I had that the mathematical description of nature called Pursuing the Disintegral is not quite as good as a Theory of Everything. A Theory of Everything is still required. —How could someone debunk the unified theory of cosmic plasma physics?

January 2018. Hao Huang (2019) anticipated by Nathan Coppedge here: Truth Psychology (Jan 2018)

- Sensitivity Theory: $s(f) >= \sqrt{degree}\,(f)$.

Philosophical Razors concept (June 10, 2018):

Philosophical Razors
What doesn't make sense to you? That there's no centralized idea repository.

What doesn't make sense to you?
Jan 2019. Leibnizian Characteristica Universalis was incomplete:

Coppedge, Characteristica Universalis
May 2, 2020

"Later note: If proportional numbers are seen as line segments, this provides a possible road to consider ratios of infinite distance as conventionally rationalizable numbers." —Guide to Mathematical Completeness

May 3, 2020

Measuring wind audibility: "I suppose it depends first on distance, then on speed, then on audibility." —If wind is converted into sounds how random or concise is the data source?

May 6, 2020

THE EQUATIONS AND SOLUTIONS TO UNSOLVED PROBLEMS

Scientific Standard of Constants: <u>Why is the 'theory of everything' paradoxical?</u>
May 24, 2020

Impossible space in physics: <u>Impossible Space Theory</u>
May 30, 2020

Prediction of westward movement of the N. American continental plates following break from Europe. Humans with greater diversity wanted to be allied with missing land-masses like penguins: <u>What level of certainty is there regarding the arrangement of continents 10s-100s of millions of years in the future?</u>
October 14, 2020

<u>Room-Temperature Superconductivity Achieved for the First Time</u> Perhaps related: <u>Unit Spectrum Theory</u> "(+/- 16 may equal microfunction theorem). Also, Unit = D + 2 + 6 + 8... Sulfur. Hydrogen + Phosphorous. 2-Hydrogen + Silicon." (Aug 21, 2020).

July 1, 2021

- Equation for energy finally accurate version of Einstein? FOR ORDINARY OBJECTS [(MIN EFF + 1) - (MAX EFF + 1)] / [0.5 (MIN EFF + MAX EFF)]. In this case + 1 would equal the potential energy of the atoms.

Nov 2, 2021

- Theory that Perpetual Motion was an Undiscovered Negative Dimension: What is Metaphorically Rolled Up at the Planck Length? —Instead of being rolled up at the Planck length, some of the dimensions are negative, 'inside-out dimensions', including impossibility [-5], coherence [-4], growth [-3], perpetual motion or standing waves [-2], and the mathematics of nature or disintegral [-1]. Otherwise it involves more than -1 dimensions. (—2021–11–02)

November 18, 2021

<u>Improved T.O.E. Technology Equations</u> Prediction of universal universal technology equations based on TOE: for example, applications to time-travel, immortality, teleportation, telekinesis, biology, chemistry, anything anything.

December 12, 2021

Anticipation by N Coppedge of using Antimatter as a unifying force to explain how atoms move without gravity: 1X Antiforce Max: (D + 2) / 2... If it is correct, and if antimatter equates with antiforce, then in theory within a vacuum and without gravity the mechanical antimatter energy adds up to no more than gravity and perhaps self-propulsion. That is, within the 3rd dimension. —How does changing light into matter and anti-matter affect the future of science and humanity?

2022

In June 2019, Nathan Coppedge wrote about The Theory of Everything formula later used in a scuplture by Fabio Zanino: Decostruzione XC | Fabio Zanino

September 23, 2022

- Nathan may have proved the occult book 'all you need is four' was wrong with the phrase: "-2 Title + 6 Categories = 4, Describes the contents of any coordinates [souls]".

...

THE EQUATIONS AND SOLUTIONS TO UNSOLVED PROBLEMS

PLACES WHERE COPPEDGE'S WORK MAY BE OWED TO OTHERS:

- Sappho, 610 BC, Attributed to Sappho: What we worship is the opposite of what we are. May have required considering that the coherentist is made of nothing, but worships rarity.
- Socrates, 350 BC, anticipates Formula tor souls (2016) with his idea that the soul is ironic.

1730: "Christian Wolff" 's *Philosophia Prima* makes a distinction between the 'special metaphysics' of 'the soul the world, and God'. In a less religious sense, Nathan Coppedge makes use of a similar distinction between the 'God-Trap, the Zero-Trap, the World-Trap, and the Philosophy-Trap' which form a basis of two categories together labeled 'Traps' which occur in some of the low-energy categories of the T.O.E. devoted to Time-Traps and 'Damages', and also medicine.

1782: A concept from Diderot (sweet of the sweet, "Suite de l'entretien précédent" D'Alembert's Dream - Wikipedia) of the sweet spot anticipates the general concept of the sweet spot of technology in Nathan's writing Classification of Black Swans

- 1905, Einstein anticipated non-absolute universal constants. Nathan, 2021–12–14: These values X 10 ^ -2 may be seen as coordinated with un-absolute universal constants. —TOE^2 Inspired by: "We see for the first time how the light is not only delayed due to a strong curvature of spacetime around the companion, but also that the light is deflected by a small angle of 0.04 degrees that we can detect. Never before has such an experiment been conducted at such a high spacetime curvature." ---Challenging Einstein's Greatest Theory in 16-Year Experiment – Theory of General Relativity Tested With Extreme Stars

- 1905, Einstein anticipated the solution to infinite pain: "The 4th dimension involves souls and nothingness. The 5th dimension involves interactions of souls and interactions of nothingness. The 5th dimension is about relationships. Love and war are at the 5th dimensional level. Advancing to the 5th dimension could require multiple souls instead of one. This could be a reason to stop the cycle towards infinite pain." —Times Nathan Coppedge Saved the World

Scarily, Vladimir Lenin had a book called: <u>Materialism and Empirio-criticism</u> (1909, predating <u>Critical Empiricism Manifesto</u> by 112 years. Katy R says Lenin gets everyone).

- Sigmund Freud (1920): Someone intelligent could achieve understanding people, understanding human psychology, understanding human passions. Later example: Examples: Someone emotional could achieve knowledge, souls, machines (2013 - 2016).
- 1940s or later, Lacan or somebody beat me to the concept that modernism is a language disorder. This involves a few of the main 'Chinese ideas' which make effective idea combinations. Though this seems rare, these ideas are potentially ahead of some of what Brian Coppedge was thinking at a later date.

Elizabeth Bishop (1979): A diagram which included Exponential Knowledge and Exponential Mechanics, anticipating the Theory of Anything and pretty much everything Nathan works on as of 2022, was inspired by: And look the last, of ghostlike-loved-houses went (Elizabeth Bishop, 1979). <u>Works Inspired by Poets</u>

- Brian Coppedge: "Divide the board. Put the best thing in each square. Treat the squares as equal, with equal weight. Improve the squares. Improve knowledge. Make new squares." —Advice to Nathan on improving his thinking

Internal limits of Mayer Humi may have helped inspire the Theory of Everything. Perhaps Humi already had the Theory: <u>Mayer Humi</u> (...)

- 1990's Nathan believes Nathan's brother Brian Coppedge thought of a kind of coherent system of inventions using this remarkably concise system for arriving at the best possible unique inventions: Come up with your own unique ideas! Major categories: Practical, Abstract, Logic, Combinations, Languages. Then apply each major category to the following list: [biggest idea] super useful, [biggest idea] applied everywhere, [biggest idea] working with it's opposite, [biggest idea] applied to a big problem, [biggest idea] involving applied communication.

1990's The basis for the Theory of Everything may be in a method by YY: "Polymathy - -> Psychology - -> Death - -> Magic" first noticed consciously by Coppedge in September 2021. It turns out this is the crudeform of the form which appears in the TOE Thermo Diagram: "Perpetual motion, Knowledge, Death Magic, Immortality"... which is a significant expansion of the Theory of Everything based originally in Thermodynamical variations. See al-

THE EQUATIONS AND SOLUTIONS TO UNSOLVED PROBLEMS

so: <u>Improved Crudeforms</u>

1995: Mergence theory likely owed to Rick McAllister, some call him a 'computer god': *Philosophy The last opportunities for philosophy to be absorbed were (currently): Newtonian science. Hegelian Liberalism. Nietzschean Psychology. Mathematical coherence. Post-60's Humanism, self-help, etc. Scientism focused on the literal view. Eclecticism including Asian philosophy, heuristics, aesthetics, and philosophical category theory (and I would argue, ultimately perpetual motion machines).* On Rick McAllister, see also: <u>Primal Efficiency</u>

- 1997: YY may have had an equation for tutors: (Abstract or Material) / Mathematical Integration of Nature. This may be seen as the foundational hint for Nathan Coppedge's abstract and material equation, and Nathan Coppedge's disintegral equation.
- 2001: Chameleon, Hypochemeoan, Amphibian. Joseph Roach may have guessed this combination with the help of a government program.
- 2001: A young female fellow-freshman at Bard College mentioned Quantum Logic in a personal conversation with a few other students:
- 1 Deductive?
- Then, does it claim to be grounded, how?
- Logical, then it must assume reasonableness.
- 2 Not deductive?
- Then your argument might be limited or unreasonable (non-logical).
- A limited or unreasonable argument doesn't provide proof. If it is non-deductive it is always possible it could be unreasonable. Even if it is reasonable, a limited argument doesn't provide enough grounding for complete reasonableness. A limited argument could be insane.
- If it is non-deductive, it is not logical, by definition. If it is logical, it must assume it's rational, but assuming it's rational doesn't provide a groundwork for being logical.
- Better guess is what you need, but it's not likely.

The same fellow student at Bard may have said (though I wonder also if I said it): "aesthetic survival = semantic omniscience" this would have placed the student 21 years ahead on the <u>Complete Genius</u>

of the Universe project, where it appears in slot (4.20).

- 2005. By 2005 or so, Michael Coppedge Nathan's Dad had built something similar to the 'Improved Wheel' calling it a 'coast bicycle' saying it was an illusion more so than a real thing.
- 2005, KATY R: Katy may have related a general formula for some of the biggest ideas: [Abstraction that represents a neutral area regarding a near-universal concept]. This formula may have lead to such ideas as Exponential Efficiency, Proportional Numbers, and the Function Spectrum. Nathan may have written a similar but slightly different rule by 2013. There also may have been some time-travel involved, so I am not 100% sure that Katy is the originator, but she was someone who had skipped 4 or 5 grades.
- 2006 Stephen Hawking's book titled 'The Theory of Everything' seems to refer to a concept called 'desicated time' that I have used in my work: "[Quantum++→] Feynmann's sum over histories to the universe, the analogue of the history of a particle is now a complete curved space-time which represents the history of the whole universe" (pp. 87). My sense was Hawking was referring to the entire universe being quantumly-historically polarized.

2006 JOB INTERVIEW: Personality Profiles Interactive Document (FOR DOWNLOAD ONLY) Google Sheets (…) Does this count as genius: Maybe not: I thought of this coherent concept of personality in 2021, whereas in my job interview at the New Haven public library around 2006, my future boss knew instantly that I was a very thoughtful person. This may have required the "Personality Profiles" unless she was faking it. The difference there is almost 15 years. Was my boss a time-traveler who had already seen my work from the future?

- 2009 (?): Title: "Brecht Corbel's view of Quantum Computing" Soul: [You might be bright to study architecture], either through conventional computers, or using logic, therefore, you might not be bright if you are not the architect. So, basically, 1. Brecht Corbel advises studying architecture as often as possible. 2. Brecht Corbel anticipates insights into logic involving conventional computers, not quantum. So, Corbel hates quantum computing because of its limitations. 3. Corbel sees the importance of designing these computers yourself if you want the IQ advantages related to them.
- 2009 I heard someone with a similar name to Ainan Cawley (IQ 263 - 349) talk for a few minutes because he was a family friend. It turns out it was not Cawley, but instead a family friend who was of Caucasian ancestry (though he did have a somewhat high IQ). I

THE EQUATIONS AND SOLUTIONS TO UNSOLVED PROBLEMS

immediately felt like I needed to write some things down even though the young person who I thought was 5 to 12 years old was using potty humor. However, I wasn't sure immediately what to write down and ended up leaving the paper blank. I suspected it had something to do with mathematics, possibly the color grey, and the eternal storm but otherwise I wasn't sure what to make of it. It occurred to me later I may have mentioned my interest in coherence theory, which I had developed since 2004.

2009 DISCUSSION WITH MY DAD: <u>Function Spectrum</u> (...) Does this count as genius? Maybe not: In 2009 or so my Dad, Michael Coppedge made reference to "Jorge Vargas's many inventions" which may have required or else inspired knowledge of the Function Spectrum. This would be 11 years in my Dad's favor. Was my Dad a time-traveler who already knew about my future work on perpetual motion theories?

2010: At least by 2010, perhaps as early as 1970, educators had beat Nathan to the concept that 'qualia come in last'. This observation became important in Coppedge's writing: <u>Core Pinnacle Technologies</u> where evidently a category similar to external topology labeled qualia became associated with the last of twelve pinnacle equations.

- 2011: Viewing it as alchemy could be a broader method than just reaching results. My Dad may have mentioned that someone at the University of Nottingham was working on something like this.

2011: Details go missing, like Nathan inspired the title for the book: <u>The Universal Computer</u> which was published as late as 2011.

- 2011 Jonathon Keats may be ahead of me on 'Smarblime Realities' by this date. I later determined this to be an advanced idea which may have been ahead of my thinking.

2013 Lawrence Paulson (https://www.cl.cam.ac.uk/~lp15/papers/Formath/Goedel-logic.pdf), anticipates my 2023 work: <u>GEOMETRIC ToE</u> See also: <u>Mathematics of Alternate Dimensions</u> (alternatives to exponential efficiency) by N. Coppedge.

- 2015: Who except time-travelers really knew about art that resembled Hyper-Cubism in the 20th century until 2015? Why do all the Juan Gris books look fairly new and crisp? And Francis Picabia seemed to appear out of nowhere around 2015. Not everyone knows that those publishing on Amazon, particularly mainstream publishers, can choose any publication date they want on an honor policy.

2017. TOE^2 (October 21, 2020). A level of the game Geometry Dash, released in 2017, anticipates the title of TOE^2 https://www.youtube.com/watch?v=ulLB504xwu0&app=desktop

Dec 2019 Lindgren and Liukkonen's formula for fixed probability of particles using Heisenberg uncertainty may anticipate Coppedge's independent observation and measurement formulas from 2020. Both use the value 0.5 <u>A new interpretation of quantum mechanics suggests that reality does not depend on the person measuring it</u>

SELF-SET PRECEDENTS

March 14, 2023: Dimensional metals. Precedent in Nathan's work mentioning 'mirror-worked metal' early 2000s. <u>Metals, Metallic Minerals Research</u>

- October 17, 2023 Nathan thought of a term 'hypochemaeon' meaning being exposed to too many chemicals, anticipated in 0 AD by inventing the word 'chameleon' by his past-life as Orchyrae of Alexandria.

OTHER PEOPLE ANTICIPATING OTHER PEOPLE'S WORK

2015, Internet-approach: "But just as DeepMind was scaling new heights, things were beginning to get complicated. In 2015, two of its earliest investors, billionaires Peter Thiel and Elon M..., symbolically turned their backs on DeepMind by funding rival startup OpenAI. That lab, subsequently bankrolled by $1 billion from Microsoft, also believed in the possibility of AGI, but it had a very different philosophy for how to get there. It wasn't as interested in games. Much of its research focused not on reinforcement learning but on unsupervised learning, a different technique that involves scraping vast quantities of data from the internet and pumping it through neural networks. As computers became more powerful and data more abundant, those techniques appeared to be making huge strides in capability." —<u>DeepMind CEO Demis Hassabis Urges Caution on AI</u>

- The joke formula for the Theory of Everything called 'everybody dies' (that formula is not mine) is anticipated by an earlier formula 'Someone is a psychopath, everybody dies' modeled on the idea that God is called 'someone'.

- Poetic license is also a concept that was once thought to be a Theory of Everything.

FOLLOW-UPS ON NATHAN COPPEDGE'S WORK:

- Between 2005 and 2011 there were numerous design repercussions following from Nathan's work in naming a few business locations and designing a sculpture and the outer body of the 'cooper minicar'. Perpetual motion applications were even considered by some including Nathan's dad, however, Nathan did not feel mature enough to pursue these ventures as he had a lot on his mind (omniscience theories and a ToE were still years away from being written up). One location was supposedly a perpetual motion escalator that would be built at an airport, and also a man was interested in testing his own version of Nathan's catspur shoes for an amusement park.

Several people have been interested in reproducing Nathan's perpetual motion experiments including Jer Ram from Texas who replicated an experiment.

Michael Y.T. Hwang mentions Nathan Coppedge's write-up of his theory.

A LIST OF COPPEDGE 'DISSES'

Nathan Coppedge gave Stephen Hawking the Theory of Everything in 2007, however Hawking did not make the formula very obvious on his gravestone except in Chinese. Nathan published a work on intuitive Chinese in 2017, a year before Hawking died. <u>Stephen Hawking | Westminster Abbey</u>
<u>Coleridge Initiative</u> (b. gates foundation)

<u>Statue of Unity</u> (tallest statue in the world, but not devoted to 'over'-unity. However it does seem to credit Nathan as being a genius if you read it in Chinese)

- U.S. House Committee on Ethics (U.S.H.C.E.) sounding similar to Ush Coppedge Esquire instead of Nathan Coppedge is Empirical (National Committee on Ethics would sound like NCE). Starting in 2011 the name was changed from Committee on Standards of Official Conduct (CSOC) which sounds like it would agree with everything.

COPPEDGE 'DISS-BACKS'

Recoilless Rifles was almost invented by Hawking: Nathan argues the concept may have been invented by Nathan when Hawking died as an intentional inspiration on advanced physics, because it seemed like Hawking would make a good brand of targeter, and it was not known if Hawking had thought of a weapon before. Nathan thought of the concept of 'recoilless weeping' which sounds like 'recoilless weapon'. Nathan argued that he had invented it and not Hawking, because Hawking was 'pretty legitimately ghosted'. That is when it was determined the recoilless rifle would probably be used in brutal murders by the Israelis who would call it the 'virtual reality gun'. Nathan cried about this, but he thought the concept was 'still useful for the sake of genius, or for lack of a better genius he added darkly'. He thought, 'It is diabolical, but I am convinced it does something for the nature of pleasure, even if it is not so easy to prove it so.'

[END]

THE EQUATIONS AND SOLUTIONS TO UNSOLVED PROBLEMS

FURTHER READING

The Scientific Theories

The Scientific Papers

100 Great Perpetual Motion Machines

50 Great Flying and Underwater Perpetual Motion Machines

Perpetual Motion Machine Designs & Theory

Perpetual Motion Physics for Non-Skeptics

Golden Ages of Philosophy

BIO

NATHAN COPPEDGE is a Philosopher, Artist, Inventor, and Poet who lives in New Haven near Yale.

www.ingramcontent.com/pod-product-compliance
Lightning Source LLC
Chambersburg PA
CBHW060824170526

45158CB00001B/72